A Professional's Guide to Systems Analysis

A Professional's Guide to Systems Analysis

Martin E. Modell

McGraw-Hill Book Company

**New York St. Louis San Francisco Auckland
Bogotá Hamburg London Madrid Mexico
Milan Montreal New Delhi Panama
Paris São Paulo Singapore
Sydney Tokyo Toronto**

Library of Congress Cataloging-in-Publication Data

Modell, Martin E.
Professional's guide to systems analysis

Includes index
1. System analysis. 2. System design. 3. Computer software—Development. I. Title.
QA76.9.S88M64 1988 004.2'1 88-8885
ISBN 0-07-042632-5

34567890 DOC/DOC 93210

ISBN 0-07-042632-5

The editors for this book were Theron Shreve and Rita T. Margolies, the designer was Naomi Auerbach, and the production supervisor was Dianne Walber. It was set in Century Schoolbook. It was composed by the McGraw-Hill Book Company Professional & Reference Division composition unit.

Printed and bound by R. R. Donnelley & Sons Company.

Contents

Illustrations

Preface

This book was developed from my lecture notes for an internal company training course: "Fundamentals of Systems Analysis." Thus it is not intended as a formal textbook or as a definitive treatment of the many topics covered, but rather as an introduction and overview for beginning analysts, and as a reference work for more experienced personnel. The material was put in written form to give students something to take back to the office and hopefully use for future reference. It is a practically oriented work and contains bits and pieces of material assembled from over twenty years of doing actual analysis on both large and small application systems.

After programming (coding) there is no more basic skill for the data processor than systems analysis. Systems analysis is the process of discovering what is current practice, what are the current problems, what is missing, and what is superfluous. Thus the reader will find many lists of questions embedded in the text. The answers to these questions define the problem to be solved—that is analysis. The solution to the problem, as defined by the user's needs and desires as well as the analyst's own evaluation of the problem, becomes the design.

Systems analysis is the process of determining what changes must be made, and where. Systems design is the process of determining how those changes are to be made. There are many ways to resolve systems problems, and they depend upon the resources and time available and upon business needs. There are many methodologies which can assist the analyst in creating an appropriate solution.

It was my intent when writing this text to give analysts some insight into how to look at the environments they are to examine and how to question what they see. Good analysts must assume nothing, but rather must look for the underlying reality of the situation, and ask questions about the nature of that reality. The analyst must ask the questions: when, where, who, what, and, most important, why.

I wish to thank all of my colleagues who reviewed this work and who made suggestions for its improvement. I also wish to thank all those who encouraged me to rework my original material into the text which follows.

M.E.M.
1987

Part

1

Tools and Techniques

I know that you believe you understand what you think I said, but I am not sure you realize that what you heard is not what I meant.

Chapter

1

What Is Systems Analysis?

CHAPTER SYNOPSIS

Systems analysis is the process of breaking down a complex problem into its component parts, examining those parts and reconstituting them into a more efficient, effective whole. Most organizational systems deal with the processing of data. This processing can be viewed as the organization's reaction to these data stimuli. The analyst's task is to identify the data stimuli, follow the processing sequences activated by the stimuli, and identify the results of the processing sequences. The analyst must also determine what if any problems exist in these processing sequences and determine, if possible, how to make them more efficient and effective.

This chapter provides definitions of a system, analysis, and systems analysis and discusses the role of the analyst in the development process.

Introduction

In most organizations, data processing management must allocate roughly 40 to 60 percent of its systems budget for the purposes of postimplementation system maintenance and enhancement. While some of these postimplementation changes are legitimately due to business changes which could not be anticipated when the system was originally designed and implemented, the majority, perhaps 80 to 90 percent, are due to incomplete or inaccurate front-end analysis in the original project.

The typical systems development project works on the 80–20 rule: 20 percent of the time is spent in analysis and 80 percent of the time in implementation. What this rule does not recognize is that after implementation, as many, if not more, resources will be allocated to the maintenance and enhancement efforts as were allocated to the original development project itself. Aside from the analysis needed to add new functionality or to change the old, a good portion of these analytical resources must be spent in determining where the original analysis was either incomplete or, worse, faulty. In effect this effort is expended in redoing or correcting work already supposedly completed.

While it may not seem feasible or practical to spend more time in the analysis phases, doing so will save time and, more important, expense in the long run. Even if the time frame for the project is short, it is feasible and practical to make the time spent doing the analysis more productive and effective. Put a different way, it is possible to make analysts more effective by preparing them for their job with better tools and techniques.

The process of systems analysis is unfortunately not one of science, but one of art. There are no hard and fast rules which, if followed, will produce the perfect analysis, although there are many guidelines which will help produce a better analysis.

There have been many books and articles written on the subject, all claiming to hold the key. These books and articles usually represent one person's, or one group's, method of analysis, a method which worked for them and which by extension they feel will work for everyone else. Many of these methods are presented as the one true way. Unfortunately, there is no one true way; there are many ways, some good, some better—the best being those that work for the particular analyst.

There are many techniques and methodologies of work organization, which if properly understood, selected, and used, will yield the desired result of providing better control over the analysis tasks. There are frameworks for multilevel analysis and design, ones for development of integrated applications within both complex and noncomplex environments, and ones which work for simple applications. There are process-driven analytical techniques and data-driven analytical techniques.

These methodologies, while they may provide structure to the analysis process and may provide forms, tables of contents, and checklists to assist the analyst, do not in most cases provide guidance in one crucial area. They assume that the analyst has the information and only needs help in structuring and documenting it. They do not explain how to get the information, and if they do, they do not explain what to do with it once it has been obtained. They describe the need for functional descriptions, but in most cases, fail to define just what a function is, or what information about the function the analyst must gather.

This book will focus on the techniques of data gathering and data evaluation—on analysis. Although analysis is normally followed directly by design, this book will not cover design. It will, however, attempt to explain what the analysis process is all about and what its goals are. It will present, in the form of lists of questions, examples, and techniques, aids which may be used in the analytical environment. These techniques may be used to examine the organizational, functional, process, and data aspects of the firm as a whole, or user applications in particular. Its aim is to provide guidelines and procedures for analysis at any organizational level, and to provide a structure for the information-gathering tasks which are necessary to build effective and efficient business systems. It is meant as a practical guide and is the result of over 20 years of work by the author as both analyst and designer.

This book will cover the various organizational areas which must be examined during the analysis process and provide a variety of methods for performing that examination. It will provide what the author hopes is a sound framework of procedures, guidelines, and detailed step-by-step tasks which can assist business analysts, systems analysts, user management, and user analysts in developing the information necessary at each phase of the analysis portion of the development life cycle.

Since all analysis needs a framework or plan for accomplishing the necessary work and since there are many frameworks or methodologies in use, this book will discuss methodology in a generalized fashion. The methodology presented is one which can be translated easily into whatever formal life cycle methodology the analyst must use. This methodological framework is provided because one must be used; however it is presented with the understanding that it is not the only valid framework and is thus only a convenience for the presentation of the rest of the material.

This is a book for both the new and the experienced analyst. Many of the techniques are well known, others are relatively new, but all are applicable to some aspect of the analyst's work. Some techniques are espoused by well known firms, but most are in the public domain. All can be used by anyone with a large number of pencils, big erasers, and lots of paper.

Since the analysis process itself differs only slightly between the manual and automated environments, it is immaterial whether the analyst reader is a data processing analyst or a business analyst. It is hoped that this book would be useful for anyone analyzing a business environment, solving a business problem, or devising a change to an existing business environment.

Why Systems Analysis?

Knowledge is the necessary prerequisite for change. Change without knowledge leads to chaos. Effective change, that is, change which works, must be made within the context of an existing environment. The impact of change must be known or reasonably estimated in advance. The more complete our knowledge, the greater our ability to predict the impact of any change. When changing a system, be it a manual or an automated system, we must understand that system in order to be able to understand what must be changed, why the changes must be made, and the best ways in which to make those changes.

Information systems, particularly business information systems, consist of two parts: (a) data and (b) the processes which acquire and manipulate that data for the benefit of the business. The analysis aims at understanding the data and the processes, and determining how both must be changed to provide greater benefit. The changes may be to the data acquired, to the processes, or, more usually, to both. Although both must normally be changed, there are those who advocate examination of processes from which data needs are determined (process-driven analysis) and those who advocate examination of data from which processing needs are determined (data-driven analysis).

The contents of this book are predicated on the assumption that both data and processes must be analyzed concurrently. That is, data needs drive processing needs, and processing needs drive data needs; thus neither can be examined without, nor to the exclusion of, the other. This type of dual-directed analysis requires a variety of tools and techniques, some aimed at data, some at the processes, and some at both. The effective analyst must select the appropriate tools and techniques according to the task at hand.

What Is a System?

Because the primary medium of the systems analyst is prose and because the intent of analysis is to examine and document, all language and terminology should be as precise as possible. This is not to say that the documentation should be pompous or stilted, but rather that all parties should agree on the meaning of what has been written. For this reason we will use the dictionary definitions as the basis for our discussion of systems, analysis, and systems analysis.

A definition

Using the dictionary definition: A *system* is a "group of interacting, interrelated, or interdependent (business functions, processes, activities or) elements forming a complex whole . . . a functionally related group of (business functions, processes, activities or) elements, for instance, a network of structures and channels, as for communications, travel, or distribution." (*The American Heritage Dictionary,* Second College Edition)

"The state or condition of harmonious, orderly interaction."

"A method; procedure."

"To formulate into or reduce to a system."

As with many business terms, the word "system" has many meanings. It means different things in different contexts. In many cases, when two people discuss a system they may be referring to different things. This is especially true when a systems analyst and a user get together. To one the system is that which resides in the computer center and consists of programs and files being processed in an automated fashion. To the other, the system is a complex of procedures, work flows, and tasks which are performed to accomplish the requirements of the business.

To the systems analyst, the system's files are tapes and disks; to the user they are papers, folders, and drawers. Even when the user and the analyst agree that the system is the automated version, their perspectives may be entirely different. To the systems analyst it consists of jobs, programs, files, and software; those things which reside in the machine. To the user the system is what appears on a report or on a computer terminal screen (Figure 1.1).

For the purposes of systems analysis, a system is a business system that is described and viewed from the user's perspective, using the user's words and the user's definitions and understandings of the meaning of those words. This, of course, implies that we as analysts must understand the user's perspective. We must also understand the user's words and the meanings which the user attributes to those words. Simply said, we must speak the language of the user, and we must, at least partially, think like a user. We must be able to "walk in the user's shoes, and sit at the user's table."

This is critical to the establishment of communication between user and analyst and to the success of the analysis project. It is certainly critical to the analyst in the accumulation of that elusive body of facts

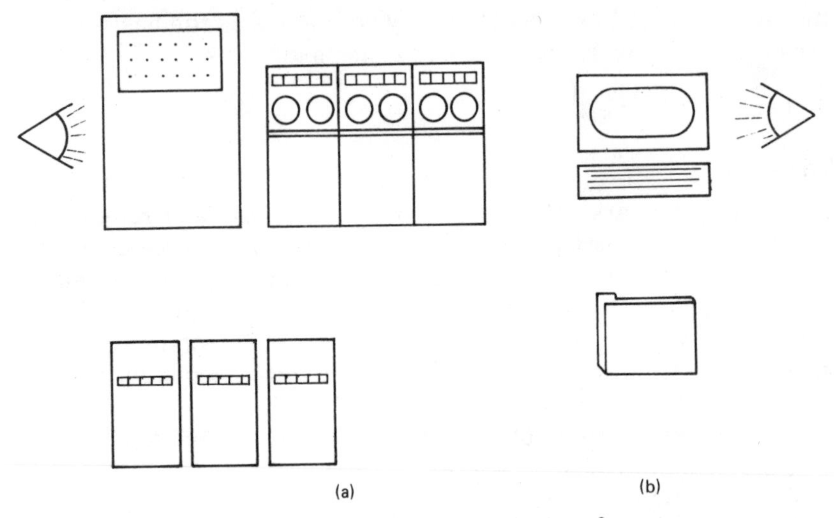

Figure 1.1 (*a*) Developer's view of a system; (*b*) user's view of a system.

known as "business knowledge." By extension, this establishment of communication with the user is a major part of the analyst's job in acquiring business knowledge and business understanding.

What Is a Systems Analyst?

A definition

In the same dictionary, *analysis* is defined as: "The separation of an intellectual or substantial whole into its constituent parts for individual study." (*The American Heritage Dictionary,* Second College Edition)

"The stated findings of such a separation or determination."

A definition

A *systems analyst* by extension is: "One who engages in the study of, and separation of, a group of interacting, interrelated, or interdependent (business functions, processes, activities or) elements forming a complex whole into its constituent parts for individual study."

The Roles of the Systems Analyst

It is as difficult to define a single role for the systems analyst as it is to define a single activity which personifies the title. This is because at various times the analyst will play some or all of the following roles.

- *Reporter.* A reporter, regardless of the medium, is a journalist whose primary task is to write, as objectively as possible, about the details and facts of an event of interest.
- *Detective.* A detective, whether official or private, is one whose primary task is to uncover the facts of an event and to determine responsibility for the event.
- *Consultant.* A consultant is one whose task is to provide assistance to a client in the form of services, advice, or guidance about a specific client-related task.
- *Diagnostician.* A diagnostician is one whose primary task is to examine the facts of an event or problem presented, and, based upon past experience or research, determine the underlying cause for that problem or event.
- *Investigator.* An investigator is similar in nature to a detective, except the scope of event under examination may be narrower, and an investigator is usually not responsible for determining responsibility.
- *Organizer.* An organizer is one who must provide sequence and order to activities and events, and must assume responsibility for the proper execution of those activities and events.
- *Puzzle solver.* The puzzle solver is one who either puts things together from component pieces or determines solutions from clues and hints.
- *Evaluator.* An evaluator is one who tests, rates, or evaluates facts for accuracy, completeness, or correctness and assigns various relative weights to them.
- *Simplifier.* A simplifier is one who looks at complex objects, events, or environments and breaks them down into less complex units. May also be one who provides easy-to-understand explanations for complex problems.
- *Indian scout.* An Indian scout is one who is usually the first on the scene and who looks for hidden dangers or for the correct path through the wilderness (of the corporate environment). The Indian scout may also be the first one to find hidden dangers and may draw the first fire.
- *Artist.* An artist is one who interprets events and environments and extracts their hidden meaning. The artist may also portray things as they are, as they seem to be, or as one wishes them to be.

The artist may not necessarily work from existent reality but may form a picture of the desired reality.

- *Sculptor.* A sculptor is an artist who works with a physical medium drawing meaningful form from raw materials or in some cases from worthless materials.

Concerns of the Systems Analyst

Analysts are key members of the development team. They are the ones who determine scope, direction, approach, and in many cases duration. Although the development team as a whole may have an administrative manager who will be responsible for the entire project, including those nonanalytical portions such as implementation and testing, personnel and resource allocation, schedule adherence, status report preparation, and in some cases making presentations, it is the analysts who run the team during the initial phases and who direct the activities through the final detail design and specification process.

Specifically the analysts may have responsibility for any one or all of the following.

1. Guiding the activities of the development team during its formative phases
2. Developing in-depth "business knowledge"
3. Understanding the user's problems from the user's viewpoint
4. Identifying and defining the business problems
5. Developing a business orientation to understand the concerns and problems of the user
6. Developing practical, cost-effective business solutions for user problems
7. Identifying, evaluating, and recommending alternatives for design and implementation
8. Identifying and obtaining resolution for all outstanding functional and user-related problems and issues which face both the user and developer prior to implementation
9. Identifying all areas where additional research is needed, performing that research, and documenting the results
10. Ensuring that the development team maintains an objective approach to the project
11. Developing clear and accurate documentation

12. Communicating the identified user's needs to the development staff in the form of clear problem definitions and accurate specifications
13. Preparing reports and presentations for user and developer management.
14. Ensuring that the organization's standards and procedures are followed during life cycle of the project
15. Ensuring that any deliverables are up to professional and industry standards

Systems Analysis as a Task of Data Processing

Most firms of any size process large volumes of information. The larger the firm, the more voluminous and complex the information it processes. That information is the lifeblood of most firms. Many firms, especially in the financial service sectors of the economy, are 100 percent information processors. Many of these firms are highly automated and could not exist for more than a day without computers. Whole industries are so automated that without computers they would cease to exist.

Because of this dependence on automated processing, most business systems are a combination of manual and automated activities, with the emphasis on the automated part. The rise in popularity of personal computers and their almost geometric increase in power and capacity, have made it economical to automate even small, limited scope tasks.

For these reasons, in most organizations, systems analysis activities are a function of the data processing organization. Thus the job title *systems analyst* is a data processing job title, and systems analysis is performed as a part of the data processing application development activity. Systems analysts are data processing people with a data processing orientation.

Typically, the data processing organization within most firms acts as a provider of automation support services, rather than as developer or producer of the products and/or services of the firm. Although the term *data processing* originally included work in both the manual and automated environments, today its role is to develop, install, maintain, and operate automated or semiautomated processing systems.

In many firms systems and operations functions are combined under one senior executive. These automated systems assist the clerical, financial, and record-keeping operations of the firm; in many instances, they replace them.

For the most part, early automated systems were oriented to the administrative functions of the firm, although as the level of automation of these functions has increased, the focus of the data processing staff has shifted to systems which assist the operational functions. Today it is hard to find an area of the firm which is not serviced by some level of automation.

Systems projects have been categorized in many ways: by size, by type, by scope, and by function. In order to understand the function of analysis, one must understand the variety of perspectives within which the analysis is performed.

Systems Projects Categorized by Scope

Systems development projects can be categorized by scope into (a) those which are usually sponsored by a single user and focus on a single application or group of applications and (b) those which are more global, integrated systems development projects covering multiple functional areas of the firm as well as multiple processing streams. These projects may also cover management information systems (MIS), decision support systems (DSS), management support systems (MSS), etc.

The former type of project is usually instigated by a user request for service, a user-recognized problem, or at times by a proposal by the data processing area in response to an analysis which uncovered either a problem or an opportunity for automation. The latter is usually begun as part of some strategic initiative of the firm's senior management, again, either from a user or from data processing.

In terms of scope and size, the former usually have a limited duration, anywhere from 6 months to 2 years. The latter have a much longer duration, sometimes extending to 3 and 5 years. These differences in levels of duration reflect the size and scope of the two different types of projects, in terms of both the analytical and the implementation efforts.

The Physician's Analogy

Regardless of their method of initiation, all projects must begin with the process of systems analysis. Systems analysis for a development project is similar in many ways to a case history developed by the physician when a new patient, or a patient who hasn't been seen in a while, comes in with a problem, or just for a routine checkup. This case history serves as the basis for the subsequent examination, diagnosis, and treatment.

The physician must be sure that the patient's history is taken in complete detail, including any known or suspected problems, habits, lifestyle, and other medically related information.

To develop a "case history" analysts must systematically and thoroughly document the "case history" of the current user environment, including the development of a dictionary or glossary of terminology and definitions as it has been presented to them and as they understand it. This documentation, which represents a view of the current user environment, when validated and agreed to by the user as accurate, constitutes the foundation for the next and all succeeding phases of both the analysis and design, and the implementation phases.

Just as the physician follows taking the case history with his or her own examination, so too the second process of systems development processes comprises a review, examination, and analysis of the findings and documentation collected during the first phase.

This review concentrates on making determinations, recommendations, and suggestions as to

Where the most profitable or cost effective opportunities for automation, or reautomation, exist

Where the manual processes of the user can be streamlined

Where the user's forms and documentation need to be changed

Where the user's procedures are deficient

Where the user's procedures need to be reworked

Whether the user's perceived problems are in fact the actual problems, or whether there is some other underlying and unsuspected problem which must be resolved

It is in this phase that the analyst seeks to address the stated requirements of the user or to determine the cause for the stated problems which the user is experiencing.

We can draw an analogy at this point to a similar and more familiar set of processes. If we are feeling ill, we visit the doctor. In the office we tell the doctor what ails us and we describe the symptoms which make us uncomfortable, or which cause us pain. The doctor then conducts a thorough physical examination and takes a medical and personal history.

The doctor then reviews the history, the results of the physical exam, and any tests which may have accompanied it. From this review he or she attempts to arrive at a diagnosis or a determination as to the cause of our problems. Many times the symptoms are a direct result of some physical problem. Sometimes the symptoms are only indirectly related to the seemingly obvious cause. In these cases, a deeper underlying problem is discovered which needs to be addressed. Once the actual problem has been identified, the doctor can prescribe a course of treatment to effect a cure, or at a minimum, relief from the pain or discomfort.

The data processing system development analysis process is very similar to the examination and diagnosis process of the medical profession. Here, the analyst seeks to analyze the symptoms and determine the actual problems which are causing discomfort to the user.

The information derived from the physical examination and the data gathering process, necessary for any analysis project, is usually termed "gathering background information" or "acquiring business knowledge." It can be obtained from experience in the firm, from the user, from user documentation, or from a variety of similar sources. The gathering of information and collecting it into a document becomes the first stage of the analysis process. One of the most common methods for gathering information is the personal interview. (See Chapter 5.)

Chapter

2

The Various Types of Systems Analysis Projects

CHAPTER SYNOPSIS

There are three types of systems projects: manual, manual to automated, and reautomation. The last, reautomation, has four subtypes: system rewrite, system redesign and redevelopment, system enhancement, and system maintenance. Each of these involves different, and yet similar, work. The work is similar in that the development activities which are involved in each follow the same general phases and approach. They are different in that the environment that the analyst must examine has substantially different characteristics.

This chapter examines each of the various types of analysis projects, along with a brief discussion of the Gibson-Nolan electronic data processing (EDP) stages of growth theory and its impact on the analysis process. In addition there is a brief discussion of the Anthony model of organizational structure.

The Three Types of Systems Analysis Projects

Analysis projects are initiated for a vast variety of reasons.

1. A change in the basic aspects of the user's functional role
2. A change in company strategic objectives

3. A need for increased performance from the automated systems
4. A need for more direct and immediate access to the firm's automated files
5. A need to upgrade the system to take advantage of more current technology
6. A need to clean up the system

The scope and magnitude of the functional and procedural changes may be fairly narrow or wide ranging. In some cases, aside from recoding the system, there may be no changes in functionality at all.

Given the variety of reasons for a project being undertaken, the starting point may also be quite different from project to project. These starting points reflect the differences in current user processing environments and the current level of user automation. Because of these differences in current user processing environments and user automation, systems projects can be categorized into three types.

1. Manual
2. Manual to automated
3. Reautomation

The last, reautomation, has four subtypes.

1. System rewrite
2. System redesign and redevelopment
3. System enhancement
4. System maintenance

From an analysis perspective, each of these types of projects involves different, and yet similar, work. The work is similar in that the development activities, which are involved in each, follow the same general phases and approach. They are different in that each of the starting or current environments that the analyst must examine have substantially different characteristics. Briefly, these six environmental types and subtypes are as follows.

Manual

From the analyst's viewpoint, this is the simplest environment in that all the components of the environment are overt. That is, they are clearly visible from observation and analysis. All work is performed by user personnel, who work directly with their files, forms, and documents. The processing of these forms and documents, the work flows, and the individual steps are easily followed.

At their core all systems projects are concerned with the examination of what are, or once were, essentially manual operations. In fact, it is helpful, regardless of the type of project, to view all the activities of the user as if they were still being performed by hand. This allows the analyst to examine in detail each task being performed, each data operation, each data movement, and each data carrier (a data carrier is a piece of paper, a form, a report, a worksheet, a transaction, etc.).

The analyst's task in the manual environment is to simplify the work flows, streamline the processes, reduce redundant processing, rearrange the tasks so as to ensure more orderly processing, and ensure that the forms, documents, and reports contain all necessary data. Each task, and each task step, must be examined to determine (a) if its execution is appropriate and (b) if it is appropriately defined, positioned, and performed.

The results of the analysis of manual systems are usually new or revised standards and procedures which clearly define the processing sequence for the task to be performed and the rules which govern their performance. In addition the analyst may develop new input forms, control procedures, monitoring procedures, and reports. The output from the analysis may also include new or revised work and data flow diagrams.

Manual to automated

Working in this type of environment differs from working in the strictly manual environment in that the analyst's task is to determine whether the manual environment, in whole or in part, can be augmented by automation, and if so, to what extent. The existing environment must be analyzed in the same manner as the purely manual, but as the analysis progresses, the analyst must also find ways of substituting automated processing for manual processing. To accomplish this, the analyst must break each process and task into its component steps and determine if the rules for performing the step lend themselves to machine automation.

The analyst's output for this type of project closely resembles that produced from the strictly manual project. However, here the analyst must also develop (a) new, input forms suitable to an automated environment, (b) file content requirements for ongoing master and transaction files, (c) report layouts, and (d) a processing flow which intermixes the original and unmodified manual processes, new manual processes, and new automated processes. The analyst must also make a determination as to the costs involved in the automation process, provide project schedules, and make hardware and software analyses and recommendations.

Reautomation

There have been many attempts to set down analytical and design methodologies for development projects in automated environments. What many of them ignore is that there are different types of automated business environments, which, while seemingly similar, must in fact be treated differently.

What distinguishes these environments is the extent and depth of automation. Early analysis methodologies were predicated on a manual environment. The aim of the analysis was to develop an automated solution to user business problems. In today's environment, most firms of any size have existing levels of automation. Many in fact have gone through two and three rounds of automation and reautomation.

Many of the existing processes and procedures are either totally automated or were developed as a result of a partial automation of the user area. Many of the forms and transaction flows within this type of environment are automated or semiautomated.

This prior automation poses a trap for the unwary analyst in that the currently used forms and documents of the business may in fact have been designed to support and accommodate an automated system. These automated systems may have been designed for the business using a level of technology which is now outdated or inefficient, or for a set of user requirements or a business environment which has since become wholly or partially obsolete. Additionally, these forms and documents are the result of some prior analyst's efforts and may not in fact reflect the natural information or data needs of the firm.

The processing flows themselves may be unnatural, to the extent that they reflect the intrusion of automated processing sequences. These flows may have been structured to accommodate the needs of the then prevalent technology rather than the needs of the business. Each of the documents, transactions and process flows must be reexamined in the light of the current business environment and the current business processing needs. They may merely need to be refurbished, or they may need to be scrapped entirely in favor of a new and more streamlined processing flow.

The analyst must look with care on batch flows, "processing windows," and transaction holding queues. These constraints may have been imposed on the processing environment by the requirements of prior automation efforts, most probably implemented under what is now an outdated, or, worse, obsolete technology. Reautomation is a major type of project which incorporates the following subcategories.

System rewrite. The "system rewrite" is one of the simplest forms of reautomation. A system rewrite usually involves very little analysis of the current business environment and thus entails very little change to that environment from the aspect of functionality or procedure. It is closely allied with system maintenance in that the goal is not to change the processing flows or to add to system functionality but simply to "clean up" the processing, streamline the existing flows, or rewrite the programs in a more up-to-date language or with more up-to-date file handling technology.

As part of the rewrite process, file layouts, report layouts, screens, and transactions may change, but usually the file and report content do not, and neither does the user processing environment. A rewrite may remove processing anomalies ("bugs"), remove unused, obsolete, or outdated code and may add *minor* new functionality to the system. The emphasis here is on minor.

System rewrites are usually done for the benefit of, and are initiated by, the data processing community, either the data processing system developers and system maintenance personnel who need a "cleaner" system to maintain, or the data processing operations personnel who have requested operational changes to streamline the processing or to make it more efficient.

System enhancement and maintenance. Although the two subcategories are usually separated, they are sufficiently similar, from an analysis viewpoint, to be examined together. Applications maintenance and enhancement projects differ from development projects in one substantial way: The analysis and design personnel assigned to these projects must also assess the impact of the proposed change on an existing system.

These maintenance or enhancement projects usually leave large parts of the base system intact. The remaining parts are either modified or "hooks" are added to the additional code which support the added functionality. The requests for maintenance or enhancement changes normally originate with the user, although they may originate with the development team itself.

There are numerous reasons for these system modification requests; among them are changes to the business environment, user-requested additional functionality, correction of erroneous processing, and user-requested refinements, or cosmetic changes to the existing system.

Those changes which originate from alterations to the business environment are the most difficult to implement, followed closely by those which add new functionality. The implementation difficulties arise because these types of changes not only require new analysis of the user area but also reanalysis of the original system design to determine where and how the changes can and should be made. Maintenance or enhancement which is necessitated by correction of erroneous processing, user-desired refinements, or other cosmetic changes usually requires little in the way of new analysis.

The analysis and redesign efforts required by business environment changes and additions of new functionality can be almost as extensive as those which were required in the original systems development. The most difficult aspects of changing an application system are those which are directed either toward changes in underlying system design and the resulting processing logic, or toward changing the structure and contents of the system files or database.

When the maintenance or enhancement project is directed at a system in a database environment and the database must be changed, the analysis must not only cover the application in question, but also any other applications which use the same database and in particular the same data records or data elements.

In some cases, the immediate enhancement or maintenance project will require data which should logically be captured by a different, unrelated application. The "chain reaction" or "cascade" of changes can increase the scope and impact of the initial request by orders of magnitude. Use of database technology encourages integrated and interdependent systems designs. The greater the integration or interdependence, the greater the potential impact of any change.

The most difficult of these data-directed changes occurs when the structural logic of the database must be modified. Even minor changes here can have major impact. The more extensive the use of the database, the more thorough and painstaking the analysis that is required. Reporting, retrieval, or other file access facilities may be severely impacted by these data structure, content, or processing logic changes.

System redesign and redevelopment. The system redesign and redevelopment project is the most comprehensive and difficult type of reautomation. It involves not only the traditional activities of the "new" development project but also the additional activities of file conversion.

The system redesign and redevelopment requires a start-from-scratch analysis, which must at the same time acknowledge the presence of the existing system. This type of project is usually undertaken when an organization "migrates" from one hardware or software technology to another.

In some cases, this "migration" is from a batch to an online environment, in others to an integrated environment, and in still others to a micro- or minicomputer environment. This "migration" usually involves new file organizations, new data and data structures, and sometimes new hardware and implementation software.

The newness of the technology requires that the existing hardware and/or software base be replaced. In some cases the business environment has changed so substantially, either for competitive or internal restructuring reasons, that the old systems can no longer service the business. The internal restructuring may have resulted from a shift from a centralized to a decentralized environment, the addition of major new lines of business, or a change in orientation of the firm, such as from an account to a customer orientation, or from a sales force–based organization to a direct-mail–marketing organization. In some more recent cases, this restructuring results from the firm's acquisition of other businesses or from the acquisition of the firm by another business. In all cases, the internal automated systems need to be redesigned and redeveloped to support the new environment.

The Organization as a Multilevel Entity

The managerial pyramid

Any organization can be viewed as a multilevel entity with each level representing a different level of control. In addition, each successively lower level has different data requirements and a different, and less extensive, view of the organization. Obviously the higher the level, the more interrelated the business functions become until, at the very top, they are viewed as one homogeneous organization with one continuous data flow.

The levels of the organization have been discussed in other contexts, but briefly for our purposes we will use the definitions of Robert Anthony in his 1965 *Harvard Business Review* article entitled "Planning and Control Systems: A Framework for Analysis." He discusses the organizational management as if it were arranged in a pyramid (see Figure 2.1). The pyramid is divided into three horizontal sections. From top to bottom these sections are labeled, respectively,

Strategic

Managerial or administrative

Operational

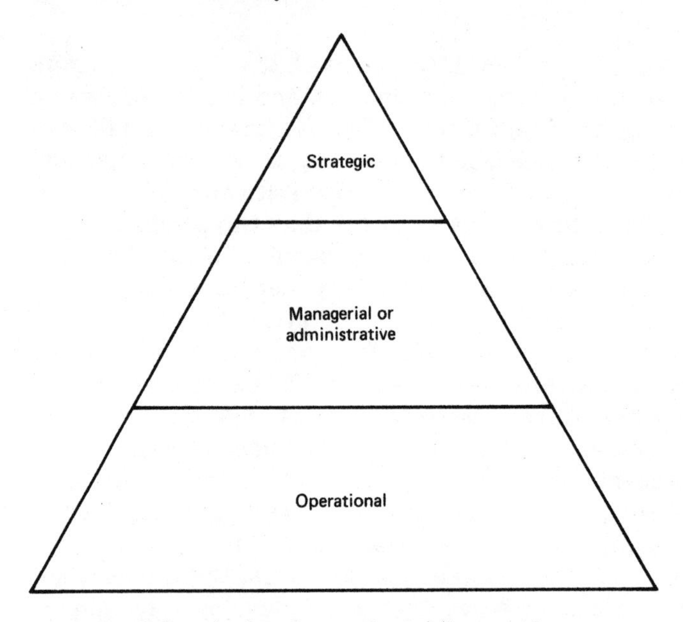

Figure 2.1 Robert Anthony's organizational pyramid.

Strategic

At the strategic level, the requirements are informational, derived from past data events and outside activities. The strategic level is the highest level of the firm and is usually populated by the most senior management. Theirs is the broadest view of the corporation. All reporting lines originate from this point. The strategic level is responsible for overall corporate policy and direction, and is primarily oriented toward functions rather than toward processes and tasks.

Strategic data are highly concentrated and usually contain little detail. In many cases data at this level may be limited to critical success factors (key numbers which indicate the operating health of the firm) or graphics which represent trends or comparison data. Much of the data at this level are financial in nature and relate directly to the critical balance sheets, and profit and loss figures of the firm. Strategic data are a mix of internally generated and externally obtained information. This information is used to compare the firm with its competitors and with the economy as a whole.

Managerial or administrative

The managerial or administrative level controls and organizes not only the company actions based upon the organizational input but also performs the supervisory activities aimed at ensuring its correct processing. It also monitors processing rates and quality. The managerial level is responsible for the tactical implementation of the policies and directions received from the strategic level. The managerial level is oriented toward functions and processes.

Managerial data are more fluid and limited than those of the operational level since the people at that level are more dependent on information than they are on data. Managerial and administrative data are almost solely derived from internal sources and reflect the operating health of the firm.

Managerial data are used to monitor day-to-day operations and may be used at either the summary or detail level. In most cases, data at this level are extracted from operational reports. Managerial data needs are not as immediate as those of the operational level.

Operational

The operational level is data and processing oriented. Its inputs are specific and derived from current data events. While management flows downward in the organization, data flows largely upward. At the operational levels all work occurs as a result of one of the following.

1. A transaction orientation
2. A response to the largely external entry of data from a data event and/or the passage of a predetermined unit of time
3. The completion of some other internal activity which resulted in a change in the status of some activity or data entity

It is the operational level which is the predominant recipient of data introduced into the organization. The operational units and their related managerial overseers are fixed in focus. Their horizons are limited to their own specific activities. Applications and systems aimed at these groups are, of necessity, also limited. The operational level is primarily oriented toward processing and tasks, rather than toward functions.

The various operational functions interact, passing data from one to the other; in the process files are built in support of the overall activity of the business and the informational needs of management. The data requirements of the operational level units, while extensive, rarely change since they are contingent upon fixed sources of input.

The Four Stages of EDP Growth

Another aid to understanding the various types of data processing projects, and thus the various types of environments in which systems analysis activities must be performed, is to understand the evolution of data processing support in the corporation. The most widely used model of this evolution is Richard Nolan's model of EDP growth stages.

In their 1974 *Harvard Business Review* article, "Managing the Four Stages of EDP Growth," Cyrus F. Gibson and Richard L. Nolan outlined a cycle of corporate data processing evolution. These stages were labeled initiation, expansion, formalization, and maturity. Today this staged view of the data processing evolution is used by the Nolan-Norton Company to analyze maturity of the data processing systems organization and to assess its developmental effectiveness (see Figure 2.2).

Most, although by no means all, organizations evolve through the four stages during their existence. In some cases, as technology changes, the firm may evolve through the cycle multiple times, once for each generation of hardware and/or software technology. The following are descriptions of each stage; however, bear in mind that these stages can be passed through multiple times.

Initiation

This first phase is typified by tentative starts at the new technology and by early pilot projects. Usually the first activities to be attempted are those which seem best suited to automation under the new technology.

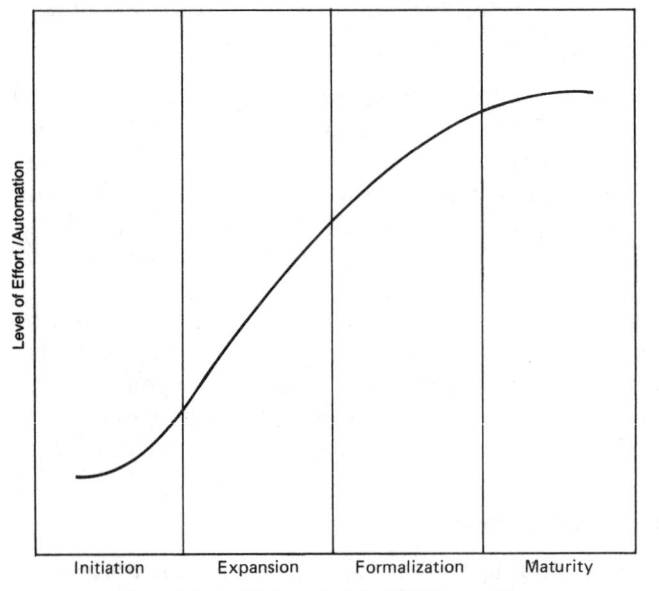

Figure 2.2 The Gibson-Nolan model of EDP growth.

Where the firm is undergoing initial automation, the projects addressed are usually activities which involve largely repetitive numeric calculations, such as payroll, general ledger, accounts receivable, or accounts payable. As the organization gains experience and confidence with the new technological tools, or with automation in general, it progresses rather quickly to the second stage, which is called "expansion."

Expansion

This stage, sometimes also called "infestation," is characterized by rapid and uncontrolled growth of applications. The data processing organization, in a rush to use the new technology, attempts to apply it to everything in sight or to place everything under a new technology base. The organization starts a large number of projects, brings in larger equipment, and hires more staff.

As these projects complete and new ones start, management begins to realize that it has spawned a monster. The data processing expenses rise more quickly than most other corporate expenses, and management begins to realize that it has no control over the data processing area. In addition it begins to realize that there is little emphasis on the traditional project and expense controls. Management at this point begins to crack down and impose standards, policies, and procedures on the data processing area.

Formalization

In the formalization phase, controls, standards, and procedures are put into place which structure the expansion phase and establish the foundation for the maturity phase. As the levels of development control are established, management sees the need for data standardization and control. The management of data becomes as important as the management of the development process. It is in the late expansion and formalization phases that the firm's need for data is replaced by its desire for information.

Unlike the earlier stages of initiation and expansion, where the impetus was for penetration of automation into user areas and the acquisition of data with little regard to coordination and integration of systems, migration to the maturity environment requires careful planning and coordination.

It is in planning this migration to a mature environment that management must resolve the organizational issues necessary for a successful effort. It is during the second and third stages of systems development that the distinction between operational and informational needs is identified and refined.

Maturity

The maturity phase is characterized by a corporate recognition of the need for integrated systems and for data and information availability. During this phase systems are usually redesigned and rewritten to use common software and common databases, new hardware or software technology, and other advanced information processing techniques.

It is the maturity phase that traditionally gives rise to the database environment and the need to integrate diverse applications into a planned architecture. It is only after the operational portfolio has been firmly established and controlled that consideration, and appreciation, of the full benefits of the technology can be realized.

Multiple passage through the model

Because the firm or parts of the firm have differing levels of automation and make use of differing types of technology, it is possible and probable that the firm will be at different stages in the model at the same time.

The growth of computer capacity and speed, and the declining costs of hardware have given the firm the capability to process data rapidly and consistently. The development of newer, faster, and cheaper machinery provides an impetus for the organization to change its processing environment. Each major change in technology can cause the firm to revert back to the first or second stages of the growth cycle. These changes, however, are usually not firmwide, and thus various parts of the firm may be at different stages of the growth model for different technologies. In some cases the same part of the firm may be at different stages of the growth curves (see Figure 2.3). Two of the latest changes, one hardware and one a combination of hardware and software, have caused this type of multistage environment in many firms. These changes are the introduction and rapid development and deployment of personal computers (microcomputers) and the introduction of viable implementations of relational database technology either via hardware (database machines) or software.

With the introduction of each new change, a new round of enthusiasm, experimentation, and familiarization is embarked upon. In many cases, new projects are initiated for the sole purpose of using the new

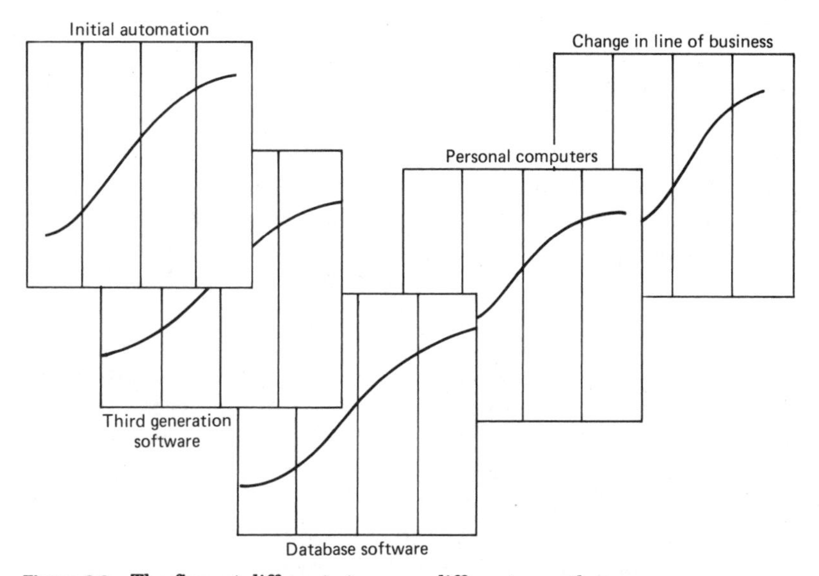

Figure 2.3 The firm at different stages on different growth curves.

technology. When we examine the growth models, however, we soon realize that controls are usually introduced late in the cycle, along with methodologies and standards. The usual reason for this is that it takes a while to determine what standards and methodologies will be needed and which will work best in the new environment. In many cases, methodologies used for older technologies are not applicable to the newer ones.

The development of methodologies and procedures, and the maturity of the technical tools and facilities provide the capability to control, manage, and maintain both the individual applications and their data. The methodologies and techniques of design, the designs themselves, and the processing may not have been changed or upgraded to fit the newer technology. This mismatch between technology and methodology causes greater disruption and lack of effectiveness in development than being in multiple growth stages. Many processing methodologies still only replicate the manual processes they have supplanted. In some cases where new technology has been the impetus for change, the new technology is employed to process systems designed or even redesigned under the older outdated technology.

Thus, we must couple appropriate methodology with technology in order to properly process and store data. Unless this is accomplished, the processed, edited, organized data will still have marginal or narrow usefulness to large segments of the firm outside the specific area of origination.

Organizations need to record data for both short- and long-term use. The systematic, short-term, accurate recording of data is basic to the successful daily operation and long-term survival of the organization. By providing a permanent record of the corporation's activities, the archiving of data sustains the auditing, statistical, forecasting, and control functions.

Usually, this data is stored in a decentralized manner which reflects the functional departmentalization of the organization. Payroll records, for instance, are normally stored in the payroll or accounting department and personnel records in the personnel department. Some records, however, are stored in more than one functional area. Copies of purchase orders, for example, might be kept in purchasing, inventory, receiving, quality control, accounting, and in the originating department itself. As each area performs its part of the processing, the base data are modified. Rarely, if ever, are all copies of the base data changed in unison; thus, to gain a complete picture of a particular transaction, one must look into the files of each area that had access to, or processed, the purchase order in some way. As a result (and to the detriment of the firm), the data in each processing area are incomplete or, worse, inaccurate. At best, it is suspect. In any case, only those areas that have copies can use the data or their part of it. Thus, the view management has of the data it receives is biased toward the area from which it was obtained. That is to say, only the data germane to a given area can be expected from that area.

Operational Versus Informational Systems

One of the primary issues facing the analyst with respect to the understanding of the users and the system which must be developed to support them is that there are two distinctly different types of functional roles and thus two distinctly different kinds of systems within the organization.

These functional roles can be identified as operational and informational (see Figure 2.4). From the analyst's perspective these two classes of systems must be viewed and treated differently.

When we look at the levels of management on the classical pyramid we can more clearly see the distinction. At the lower level, we have the operational units. Here the functions are vertical and the focus is limited to a specific organizational segment. Systems at this level are developed on a client-specific basis and are used to facilitate and control the day-to-day business of the firm. It is at this level that the bulk of the corporate portfolio of systems is developed. These systems are customized to the needs of the user client and are usually under the control of that user.

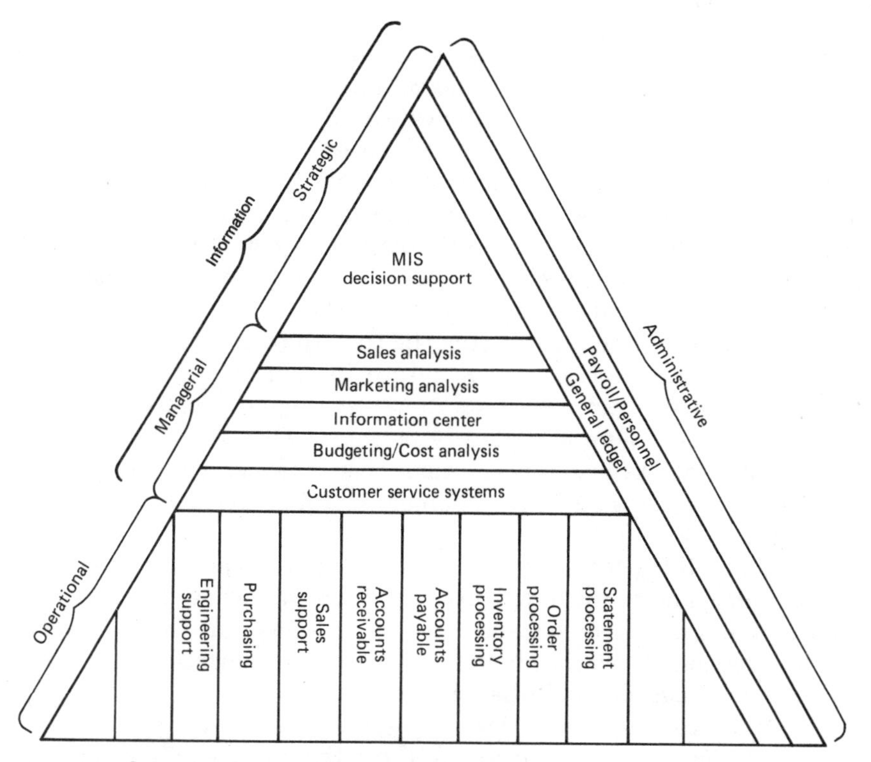

Figure 2.4 Operational versus informational systems.

In an organization where the systems portfolio is well developed but in a second or third phase of growth, management's informational needs are usually satisfied by reports which are little more than summaries of reports originally developed for operational people. These reports are extracted from the operational files.

The primary emphasis in the expansion stage is in the operational environment. Management's need for cross-unit information during this phase is usually satisfied by multiples of these summary reports, each slightly different, each representing the unit, time frames, and specific definitions of their originating organizational units.

Operational systems are those systems which support those at the lowest level of the pyramid. They are characterized as being transaction based and cyclically processed; they are usually batch oriented and operated in a current time frame. That is, the transactions are accumulated and processed on a periodic basis. The files created from those transactions represent the accumulation period and are designed for expediency of processing rather than, necessarily, for the production of information.

Operational systems are built on a function-by-function basis or functional-collection–by–functional-collection basis, and each system-supported function is traditionally called an application.

Informational systems are broader based, more horizontal in nature, and usually arise from the operational files of the firm. While there are "applications" within the informational systems, these are reportive in nature rather than processing in nature. Existing data are arranged and ordered to provide the control, coordination, and planning functions with views of the business.

These systems are retrospective in that they are concerned with what was and are projective in nature in that they project future trends from past events. The data in informational systems tend to be less precise and more statistically oriented. That is, these systems tend to look at the whole population, or samples or segments of the population, rather than individual members or occurrences of the population.

There is a third category of systems which are both horizontal (informational) and vertical (operational) in nature. They can be classified as administrative systems. In this category fall those systems which are usually overhead, and support the organization as a functioning unit regardless of what the business does.

The most common examples of these administrative systems are payroll, human resources (personnel), and the myriad of financial systems (general ledger and budgeting). The systems support the managerial planning and control needs of the firm rather than some specific aspect of the business of the firm.

As the operational and administrative functional needs have been fulfilled, automation has even penetrated the managerial and executive functions with decision support, marketing, planning, forecasting, monitoring, control, and management information system applications.

Generally speaking, these automated application systems are designed to accomplish one or more of the following.

1. Replace well defined, repetitive, manual tasks
2. Support existing work flows
3. Provide information to support the monitoring of business operations
4. Support the management decision-making process

There are exceptions to the above generalizations. These usually occur in organizations which

1. Produce products or provide services which are either in whole or in part creatures of data processing

2. Provide data processing services for hire
3. Develop data processing systems for sale, lease, or franchise

For the purposes of our discussion, it is immaterial whether the system is being developed in an organization where data processing falls into the former or the latter role. The principles and guidelines laid down here apply to the general processes of applications systems analysis, design, and development, and thus apply equally to both. Regardless of the purpose of the system to be developed, the process of analysis is the same; what differs is the application to be analyzed.

Chapter

3

The Systems Life Cycle Methodology

CHAPTER SYNOPSIS

A methodology should provide a framework or procedure within which the analyst can systematically and comprehensively investigate a business or business area, document the findings developed from that investigation, draw conclusions from those findings, and develop recommendations based upon those conclusions. In many organizations this framework has been formalized into a standardized set of steps or phases which defines both the sequence of the phases and the documentation deliverables which the firm deems necessary to the accomplishment of the desired tasks in the most efficient and practical manner.

This chapter discusses the need for a methodology and the generic phases which such a methodology should contain.

What Is a Methodology?

In order to accomplish any given set of tasks effectively, one must have a work plan. Without a work plan, activities are performed in a haphazard manner and with little, if any, coordination. The end result is that the various pieces rarely fit together into a cohesive whole, and worse yet the finished product rarely meets the initial specifications. In some cases because of the lack of a work plan, there are no initial specifications. The work plans for systems development are called methodologies.

A definition

A *method* is "a means or manner of procedure, a regular and systematic way of accomplishing something. An orderly and systematic arrangement. Procedures according to a detailed, logically ordered plan."

A definition

A methodology is "the system of principles, practices, and procedures applied to a specific branch of knowledge." (*The American Heritage Dictionary,* Second College Edition)

The Purpose of a Methodology

Methodologies, specifically data processing development project methodologies, provide a framework or procedure within which the analytical tasks can be performed. Most methodologies cover the entire span of development activities from project initiation through postimplementation review.

Depending upon the authors and the objectives, these methodologies will be either very general or very detailed. The very general ones provide a framework, and leave the specifics to the development teams. At the other end of the spectrum are those which specify each detailed task and each detailed deliverable or work product.

Aside from the development methodologies, there are others which specify the manner in which specific analytical or implementation tasks are to be performed.

Development methodologies have in common a general preference for a top-down approach. That is they begin with functional analysis (a very general view of the firm, or area of the firm) and proceed through process and task analysis (a very detailed view of specific areas). Most specify the content of the deliverables and most include other project definition, management, and control information as well. These project-related items include checkpoints, walk-through, quality reviews, management reviews, funding reviews, and large numbers of sign-offs (signatures of concerned parties, usually development management and user management).

The specific methodology followed by the firm may be home-grown or purchased from any one of a large number of firms who specialize in the development of methodologies. The methodology may be specific to a batch environment, an on-line environment, or a database environment. Methodologies exist which have been tailored to specific development products [i.e., specific database management systems (DBMS) products].

The source of the methodology is not important; what is important is that one exists, and that it be followed. From the perspective of the analytical process, a methodology provides for, and ensures that, the analyst will

1. Systematically and comprehensively investigate a business, or business area
2. Completely and accurately document the findings developed from that investigation
3. Draw supportable conclusions from those findings
4. Develop meaningful recommendations based upon those conclusions
5. In the case of analysis conducted as part of a data processing development project, design a system (which may incorporate both manual and automated procedures) which accomplishes the necessary and desired tasks in the most efficient and practical manner

The Need for a Methodology

Any methodology, regardless of which one is used, ensures that

1. All necessary activities are accomplished in the correct or desired sequence.
2. The documentation developed as a result of any given project is consistent and comparable with the documentation developed from any other project.
3. The documentation developed as a result of a given project contains adequate, understandable information.
4. Appropriate reviews and sign-offs are obtained at the appropriate points in the project.

Methodology

Generally speaking the analytical portions of a methodology must include and provide for

Business organizational analysis

Business function analysis

Business process and activity analysis

Business data analysis

This analytical portion forms the first and most critical phase of the development project. In many cases these phases are the project itself since the information developed may show that no further work is necessary, feasible, or desirable. In all cases, the results of the analysis phases determine (a) if there is a problem to be addressed, (b) if there is a feasible solution to the problem, and (c) if developing a solution to the problem is cost beneficial to the user and to the firm as a whole.

Life Cycle Phases

Assuming that the project proceeds in a normal and orderly fashion, it can be expected to follow the following general phases.

Project initiation

General business analysis

Detailed business analysis

Problem identification and evaluation

Development of a proposed general business system design

Development of a proposed detailed business system design

Development of procedural solution specifications

Implementation of the procedural solutions

Testing of the procedural solutions

Implementation of the procedural solutions into the normal business processing schedule (production)

Postimplementation review of results

Although the preceding list contains many project phases, it can be simplified to three.

Analysis

Design

Implementation

These three are bracketed by project initiation and by project conclusion and review. Additionally, all five activities include the administrative tasks of planning, scheduling, and control (see Figure 3.1).

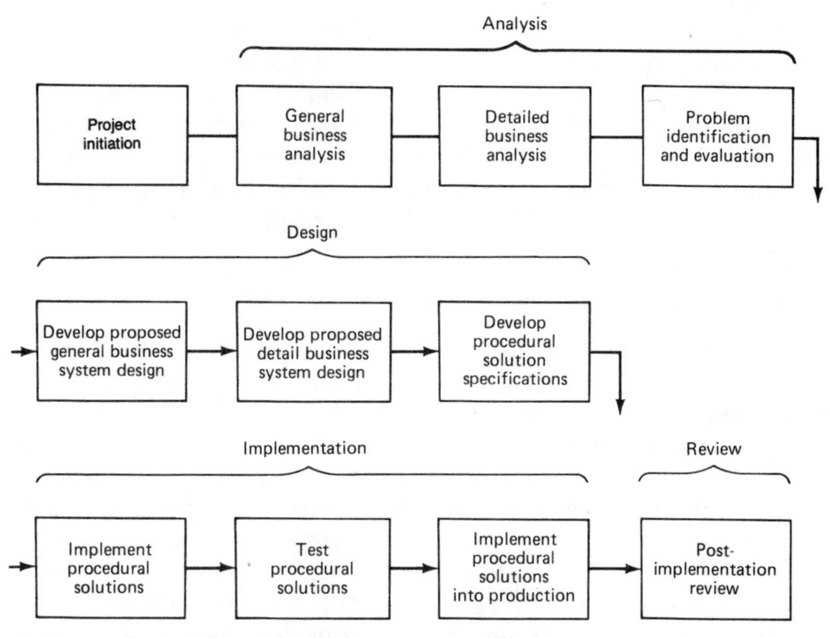

Figure 3.1 General flow of development project tasks.

Since this book is about the process of analysis, it is appropriate that we examine the portions of the life cycle associated with analysis.

Project Initiation

Normally a project is initiated by a user requesting service, by the data processing development staff, by the data processing maintenance staff, or by the data processing operations staff. The initiation process may be formal or informal. The informal route (usually verbal) is usually followed up by a more formal request, usually a memorandum or internal form. Either should state the name of the requester, the nature of the request (either the nature of the business problem to be resolved or the type of service required), the reason for the request, and the time frame within which the requester would like the service. The request should also include appropriate authorization signatures, appropriate funding information, and any priority ratings which the request may have been assigned.

It is strongly advised that no project be initiated without this information. Although the exact structure of the information may vary, it is needed for performing two vital functions: (a) identification of the user of record who will both fund and sign off on the completed project and (b) setting the initial parameters for the project.

The project initiation request sets the tone and scope of the project. Projects can be targeted at any level of the organization; thus the analyst's first task must be to determine the scope and extent of environment to be analyzed, and the number of levels of analysis which must be undertaken to achieve the desired results.

Multiple Levels of Analysis

Given Robert Anthony's multilevel view of the organization (see Chapter 2 for a discussion of Anthony's model) and an understanding of the differing viewpoints and systems needs of each level within the organization, there is an implication that there are different types of analysis and perhaps even multiple levels of analytical activity. Any methodology must recognize this and allow for analysis with a multilevel approach. Although there can be many iterations of analysis, for practical purposes we usually restrict ourselves to three. (See Figure 3.2.)

The first is identified as the *general or business environmental analysis level*, and it concentrates on the firmwide functions, processes, and data models and has been called by some "enterprise analysis." This type of analytical project has the widest scope in terms of numbers of functions covered and looks at the highest levels of the corpo-

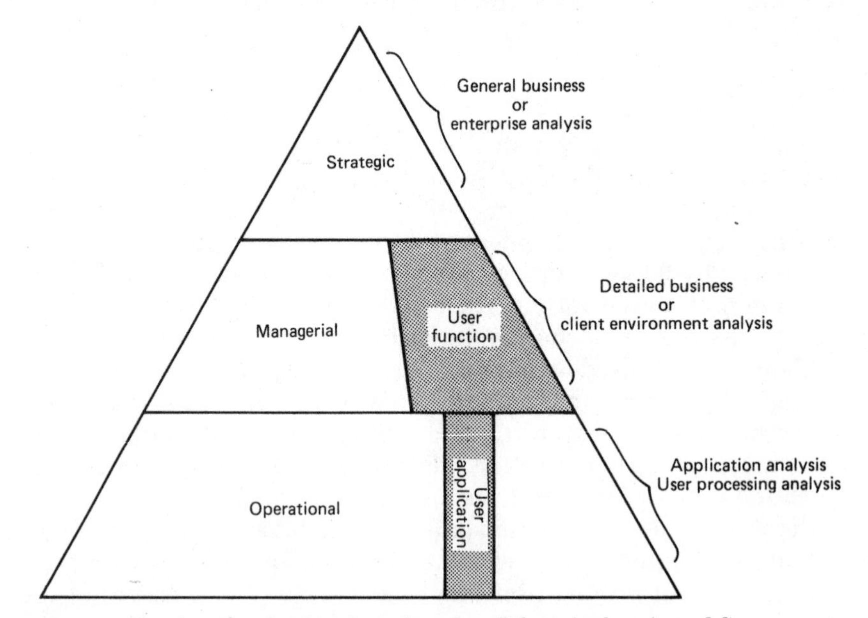

Figure 3.2 The three levels of analysis (based on Robert Anthony's model).

ration. This level corresponds to the strategic level of the management pyramid.

The second is identified as *detailed business or client environmental analysis* and concentrates on the functions, processes, and data of the individual client or user functional areas. This type of project is narrower in scope than an enterprise level analysis and may be limited to a single high-level function, such as human resources, finance, operations, marketing, etc. This type of project corresponds to the managerial or administrative level of the managerial pyramid.

The third is the most detailed and can be considered as the *application* level which addresses the design of the user processing systems. This type of project has the narrowest scope and usually concentrates on a single user functional area. It is within this type of project that individual tasks are addressed. This type of project corresponds to the operational level of the managerial pyramid.

Regardless of the level being addressed or the methodology employed, each analytical project should follow the three-phase approach.

Three-Phase Approach to Analysis

Phase 1—current environment analysis

This phase looks at the current environment from a functional, process, and data perspective. The aim here is to document the existing user functions, processes, activities, and data (Figure 3.3). The activities for this phase are

1. Current function analysis
2. Development of the current function model
3. Current process and activity analysis
4. Development of the current process model
5. Current data source and usage analysis
6. Current data analysis
7. Development of the current data model

Phase 2—problem identification and analysis

This phase examines the results of the current environment analysis and, using both the narratives and models, identifies the problems which exist as the result of misplaced functions; split processes or

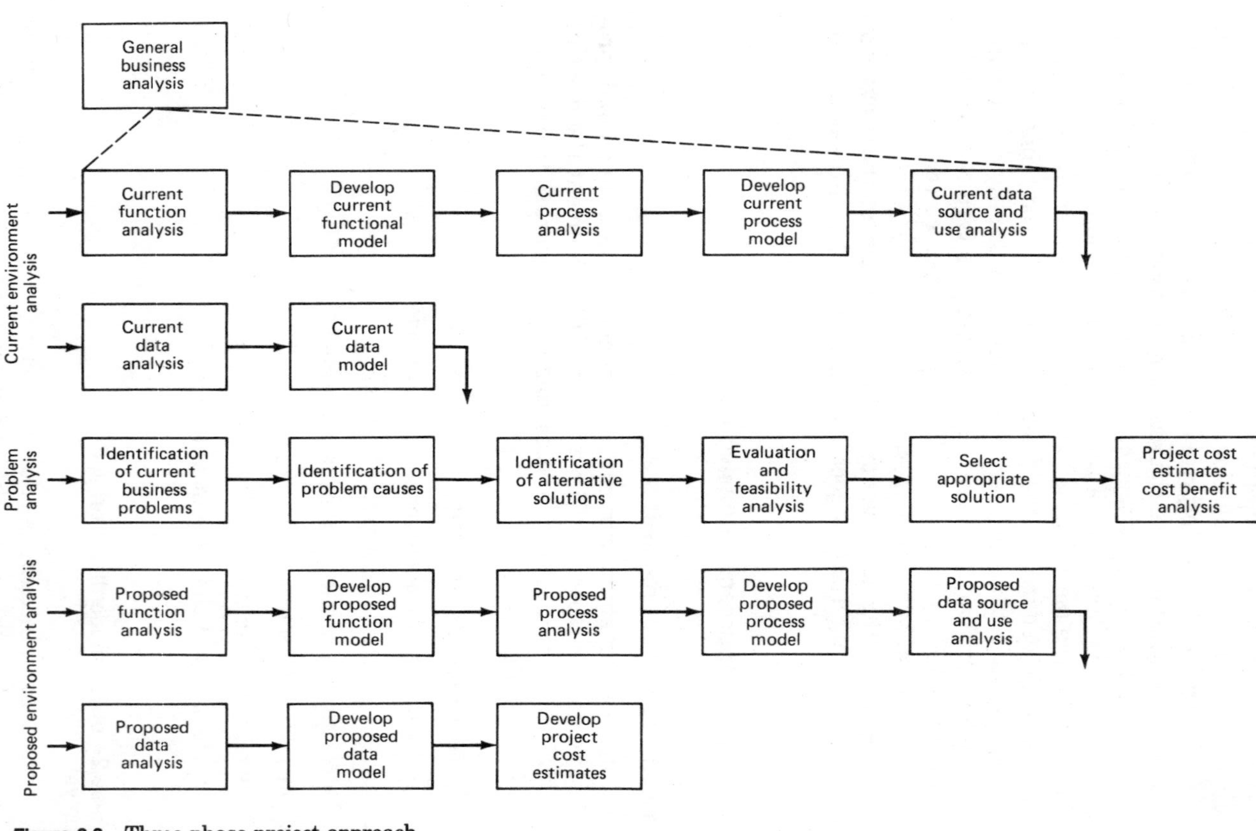

Figure 3.3 Three-phase project approach.

functions; convoluted, or broken, data flows; missing data; redundant or incomplete processing; and nonaddressed automation opportunities. The activities for this phase are

1. Identification of current environment problems
2. Identification of problem causes
3. Identification of alternative solutions
4. Evaluation and feasibility analysis of each solution
5. Selection and recommendation of most practical and appropriate solution
6. Project cost estimation and cost-benefit analysis

Phase 3—proposed environment analysis

This phase uses the results from phases 1 and 2 to devise and design a proposed environment for the user. This proposed design presents the user with a revised function model, a revised process model, and a revised data model. It is this phase which is the final output of the analysis process. The activities for this phase are

1. Proposed function analysis
2. Development of the proposed function model
3. Proposed process and activity analysis
4. Development of the proposed process model
5. Proposed data source and usage analysis
6. Proposed data analysis
7. Development of the proposed data model
8. Development of project implementation cost projections

Cost-Benefit Analysis

Each phase of a development project will normally require the expenditure of a certain level of resources, both in terms of personnel and other assets. In effect the firm is buying a product: the completed work and the accompanying documents from each phase. In order for user and data processing management to assess the relative value of each of these products in advance of their being contracted for, many firms require a separate justification document, called a cost-benefit analysis. This document provides, sometimes in estimate form, the costs for producing the product and the benefits which the firm can expect to accrue as a result of that effort.

In the early stages the project has little form or definition, thus early cost-benefit documents are predominantly based upon estimates. As more and more information is developed these cost-benefit documents become more and more detailed and more and more accurate.

Cost-benefit analysis (CBA) is used in two ways. The first is to develop estimates for project budgetary planning and, later, to apply actuals for project monitoring purposes. This is usually done at each phase of the development project and allows data processing (DP) management to assess whether the costs of the project can be justified by the potential benefits to the user and the firm.

The second is to develop user costs and benefits which will result if the project is implemented. This allows user management to plan future budgets and to estimate when the costs will be incurred. Although the costs in the first CBA contain only project costs, the costs in all succeeding analyses should also contain separate projected user costs for implementation and ongoing operations.

The first cost-benefit analysis is usually performed at the project initiation phase and is used by user and DP management to allocate a budget for the first phase. At the end of the problem identification and analysis phase a second cost-benefit analysis is normally generated. Since work has already been performed and the scope and direction of the project are fairly clear, this document can be based upon concrete information as to costs and benefits. The development alternative selected will allow the cost of hardware and software to be added, as well as the costs of user documentation and training to be estimated. At the end of the proposed environment analysis phase, user costs should be well defined and fairly accurate numbers generated.

During the problem identification and analysis phase, it is usually necessary to perform a separate CBA for each alternative being evaluated. These alternative CBA documents can assist in making the selection between multiple viable alternatives.

Project Development Costs

All costs which can be expected as a result of this unit of work should be estimated and itemized. Cost figures are usually generated by the project manager and the systems analyst assigned to the project. Generally, these costs will be broken out into one or more of the following categories.

1. *Direct personnel costs.* These costs are developed from the project

plan for this phase. (See Chapter 4 for a discussion of project plans.) From that plan, the number of people at each level and the number of periods (hours, days, weeks, months, etc.) that each person will be working on the project should be extracted. A per period cost should be assigned on any one of the following bases.

a. Standard cost. Applicable to all personnel at all levels

b. Level-by-level costs. Applicable to all personnel at each given level

c. Per-person costs. Based upon the compensation for each person (this is only used where there are no company policy restrictions on publication of personnel compensation)

The per-period cost for each person should be multiplied by the number of periods each person is expected to work on the project, and the total cost for all assigned personnel should be determined. For the best results, each person, the person's level, the costs assigned to that person, and the number of periods the person is expected to work on the project should be detailed, either in the body of the analysis or as a separate appendix. When developing these costs, be sure to include any benefits, additions, payroll added costs, and vacation and holiday costs to be assigned to the project.

2. *Indirect personnel costs.* These costs include project management, staff, and administrative services which may be needed to support the project personnel. For example, the cost related to project managers, area managers, secretaries, and file clerks, are included in this category. These costs may be calculated in the same manner as direct costs: that is, they may be estimated or taken as a percentage of the total development department administrative costs.
3. *Travel costs.* If the project requires personnel to travel between locations or if personnel must be transported to a central location, all costs of travel, accommodations, board, entertainment, and other travel-related expenses should be estimated. In lieu of an estimate, a budgetary maximum may be used.
4. *Printing and graphics costs.* The costs for development of charts, graphs, and diagrams, and for printing the final report (usually in multiple copies) should be estimated.
5. *Space costs.* If staff must be assembled from diverse locations, it may be necessary to assign office space for them during the project's duration. These costs should be calculated from the firm's space budget costs and added to the project cost totals.
6. *Machine costs.* These costs include estimates of the costs of any computer time necessary to run analyses, simulations, print machine-stored documentation, analyze machine-stored files, ana-

lyze existing user applications, create duplicate copies of user reports, etc.

7. *Supply costs.* These costs include paper, copier supplies, binders and covers for reports, pencils, pens, other stationery supplies, and any special materials which may be necessary for the project.
8. *Training costs.* These costs include any special training required by project personnel in terms of familiarization with specific business functions, special software, special machinery, general education, etc.
9. *Special software costs.* These costs include the acquisition of any specialized software for either mainframe, mini-, or microcomputers which will be necessary for the project.

All costs in each of these categories should be totaled; the sum becomes the total project cost.

User and Project Implementation Costs

Aside from the costs to be incurred by the project itself, there may be ongoing costs that will be incurred when the project is completed and implemented. These costs fall into the same basic categories as project costs, except that the staff being costed is the user area staff, etc. In addition to the above costs, the following might be included.

1. *Special user training costs.* Includes all costs for training user personnel in the use of the new system, procedures, etc.
2. *Special or new user documentation and user forms.* Includes all new forms and documents, manuals, procedural documents, etc., which might be needed by the user as a result of the new system being implemented.
3. *Special user hardware.* Includes user-owned computers, terminals, printers, personal computers, communications devices, security devices, etc.
4. *Special user software license or acquisition costs.* Includes additional copies of PC software, multiple site licenses for mainframe and minicomputer locations, ongoing training costs.

Benefits

Benefits are usually more difficult to quantify, and most benefits are shown in estimated form, especially during the early stages of a project. In many cases, these benefits numbers are obtained from the user requesting the project. They are determined by quantifying the costs of the existing environment and problems, or by estimating the savings which would result from rectifying these problems. Benefits are usually broken out into two subcategories—direct benefits and indirect benefits.

Direct benefits

Direct benefits are those whose value can be calculated or estimated in monetary terms. These include savings or reductions in

1. *Direct or indirect personnel costs.* Calculated in the same manner as development direct and indirect personnel costs. These benefits are obtained from the reduced level of effort or reduced level of staff which can be expected if the project is successful.
2. *Supply costs.* Calculated by estimating the reductions in supplies usage (paper, stationery, etc.) which can be expected if the project is successful.
3. *Machine costs.* Calculated by estimating the reduction in computer usage expected if the project is successful. This benefit area may also include reduced rental or lease costs on machinery being upgraded as a result of this project.
4. *Travel costs.* Calculated by estimating the reduction in travel and entertainment costs which can be expected if the project is successful.
5. *Interest or penalty costs.* Calculated by estimating the reduction in interest on money borrowed to finance inefficient operations, or penalties incurred for late payments, etc.

The above benefits are estimates of those reductions of such costs which might be achieved through more efficient processing if the project is successful.

In some cases the benefits may be achieved through actual increases in revenue, faster collections, more rapid turnover, or increased productivity. Some examples of benefits to be looked for, and estimated, in this area are

1. More rapid processing of customer invoices decreases the time between shipment and billing, thus the receivables period decreases and cash flow increases
2. More rapid identification and payment of vendor invoices offering discounts allow the firm to reduce its purchasing costs

3. Faster access to information allows employees to process more transactions in the same period of time.
4. Reductions in work due to a reduced number of processing steps or to simplified procedures
5. Improved editing or validation procedures which reduce errors.

Indirect benefits

Indirect benefits are those which can't be quantified, or assigned a monetary value, but which nonetheless result in desirable outcomes from the project. These might include

1. Access to previously inaccessible or missing information
2. Faster access to information
3. More flexible processing, or additional processing capability or ease
4. Additional functionality
5. Standardization of information
6. Improved communications or reporting procedures
7. Improved, clearer, or more accurate reports or screens
8. Reports providing more detail, or reports removing unneeded detail
9. Improved understanding of the basic functional or processing interactions within the firm
10. Improved documentation of the basic functions and processing within the firm

A cost-benefit analysis may be as short as a single page or may cover many pages. In format it is similar in style to a standard budget, with costs being equated to the expense side and benefits being equated to the income side. All cost items should be subtotaled by category and an overall total cost figure computed. All benefits should be subtotaled by category and a total benefit figure computed. In more complete cost-benefit analyses, both the costs and the benefits would be "spread" across the project or phase time frame.

Chapter

4

Project Planning

CHAPTER SYNOPSIS

Once the project initiation phase is completed, the project team, usually the project manager and the analysts in the early stages, must determine the scope of the effort necessary to accomplish the necessary tasks. There are many methods available for accomplishing this; many of them use graphics of varying types, but all require the same basic information.

This chapter discusses the information which must be assembled to construct a project plan and the steps necessary to construct a graphic version of a plan in either PERT, PERT/CPM, or Gantt form.

Project Planning

Once the project initiation phase is completed, the project team, usually the project manager and the analysts in the early stages, must determine the scope of the effort necessary to accomplish the necessary tasks. There are many methods available for accomplishing this planning process; many of them use graphics of varying types, but all require the same basic information.

1. Project start date
2. Project completion date
3. A list of tasks in the order in which they must be accomplished
4. An estimate of the personnel necessary to accomplish each task
5. An estimate of the personnel available to accomplish each task
6. Skill level necessary to perform each task

7. Task dependencies
   a. Which tasks can be performed in parallel
   b. Which tasks require the completion of other tasks before they can start
8. Project control, or review points
9. Project cost estimation and cost-benefit analysis

Planning Methods

There are various methods for generating the project plan. Some of the most widely used are discussed below.

PERT—project evaluation and review technique

PERT charts depict task, duration, and dependency information. Each chart starts with an initiation node from which the first task, or tasks, originates (Figure 4.1). If multiple tasks can begin at the same time, they are all started from the node and branch, or fork out from the start point.

Each task is represented by a line which states its name or other identifier, its duration, the number of people assigned to it, and in some cases the initials of the personnel assigned. The other end of the task line is terminated by another node which identifies the start of another task, or the beginning of any slack time. Slack time is waiting time between tasks.

Each task is connected to its successor tasks in this manner forming a network of nodes and connecting lines. The chart is complete when all final tasks come together at the completion node. Where slack time exists between the end of one task and the start of another, the usual method is to draw a broken, or dotted line between the end of the first task and the start of the next dependent task.

A PERT chart (Figure 4.2) may have multiple parallel or interconnect-

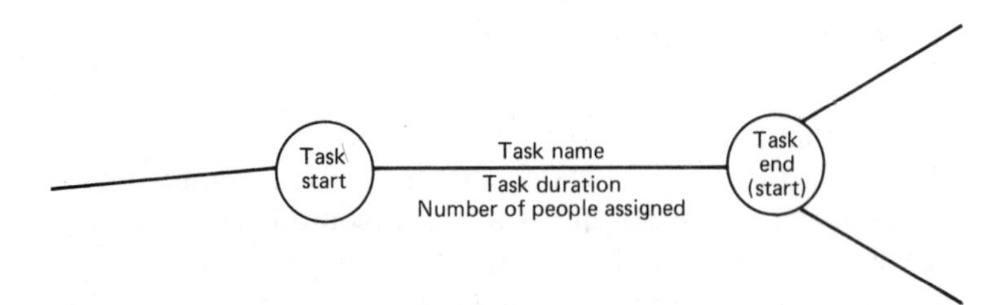

Figure 4.1 Single task line on a PERT chart.

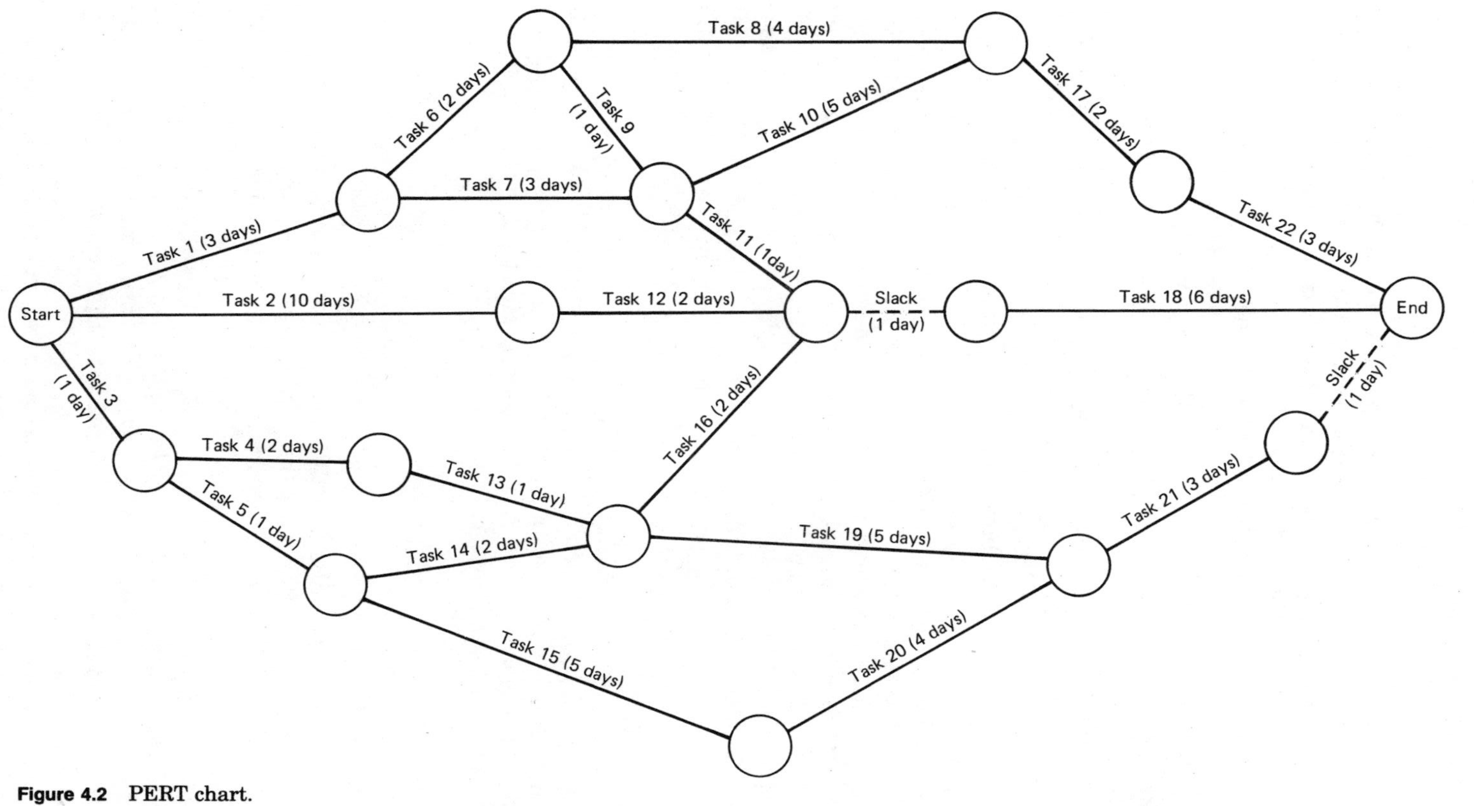

Figure 4.2 PERT chart.

ing networks of tasks. Where the project has scheduled milestones, checkpoints, or review points (all of which are highly recommended in any project schedule), the PERT chart will note that all tasks up to that point terminate at the review node.

It should be noted at this point that project review, sign-offs, user reviews, etc., all take time. This time should never be underestimated when drawing up the project plan. It is not unusual for a review to take a week or even two. Obtaining management and user sign-offs may take even longer.

When drawing up the plan, be sure to include tasks for documentation writing, documentation editing, project report writing and editing, and report reproduction. These tasks are usually time consuming so don't underestimate how long it will take to complete them.

PERT charts are usually drawn on ruled paper with the horizontal axis indicating time period divisions in days, weeks, months, etc. Although it is possible to draw a PERT chart for an entire project, the usual practice is to break the plans into smaller, more meaningful parts. This is very helpful if the chart has to be redrawn for any reason, such as skipped or incorrectly estimated tasks.

Many PERT charts terminate at the major review points, such as at the end of the analysis stage or even at the end of the feasibility analysis. Many organizations include funding reviews in the project life cycle. Where this is the case, each chart terminates in a funding review node.

Funding reviews can affect a project in that they may either increase funding, in which case more people may be made available, or they may decrease funding, in which case fewer people may be available. Obviously more or less people will affect the length of time it takes to complete a project.

CPM—critical path method

CPM charts are similar to PERT charts and are sometimes known as PERT/CPM. In a CPM chart the critical path is indicated (Figures 4.3 and 4.4). A critical path consists of that set of dependent tasks (each dependent upon the preceding one) which together take the longest time to complete. Although it is not normally done, a CPM chart can define multiple, equally critical paths. Tasks which fall on the critical path should be noted in some way, so that they may be given special attention. One way is to draw critical path tasks with a double line, instead of a single line.

Tasks which fall on the critical path should receive special attention, both by the project manager and by the personnel assigned to

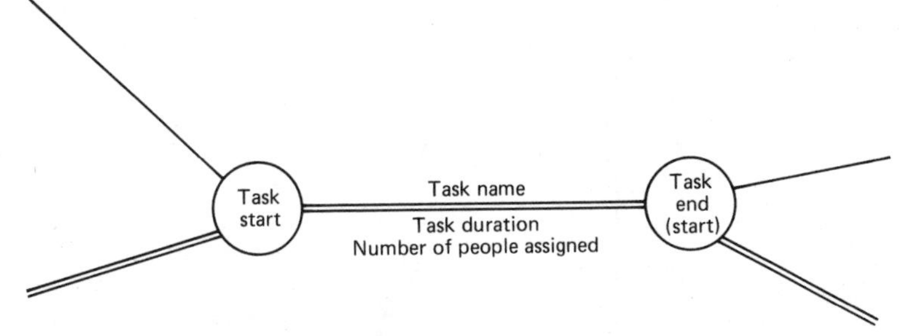

Figure 4.3 Single taskline on critical path.

them. It is possible that the critical path for any given project may shift as the project progresses. This can happen when tasks either fall behind or tasks are completed ahead of schedule, causing other tasks which may still be on schedule to fall on the new critical path.

Gantt chart

A Gantt chart (Figure 4.5) is a matrix which lists on the vertical axis all tasks to be performed. Each row contains a single task identification which usually consists of a task number and name. The horizontal axis is headed by columns indicating estimated task duration, skill level needed to perform the task, person assigned to the task, followed by one column for each period in the project's duration. Each period may be expressed in hours, days, weeks, months, etc. In some cases it may be necessary to label the period columns as period 1, period 2, and so on.

The graphics portion of the Gantt chart consists of a horizontal bar for each task connecting the period start and period ending columns. A set of markers is usually used to indicate estimated start and end and actual start and end. Each bar is on a separate line, and each person assigned to the task is on a separate line (Figure 4.6). In many cases when this type of project plan is used, a blank row is left between tasks. When the project is underway, this row is used to indicate progress, indicated by a second bar which starts in the period column when the task is actually started and continues until the task is actually completed. Comparison between estimated start and end and actual start and end should indicate project status on a task-by-task basis.

Variants of this method include a lower chart which shows personnel allocations on a person-by-person basis. For this section the vertical axis contains the name of the people assigned to the project, and the columns indicating task duration are left blank as is the column indicating person assigned. The graphics consist of the same bar no

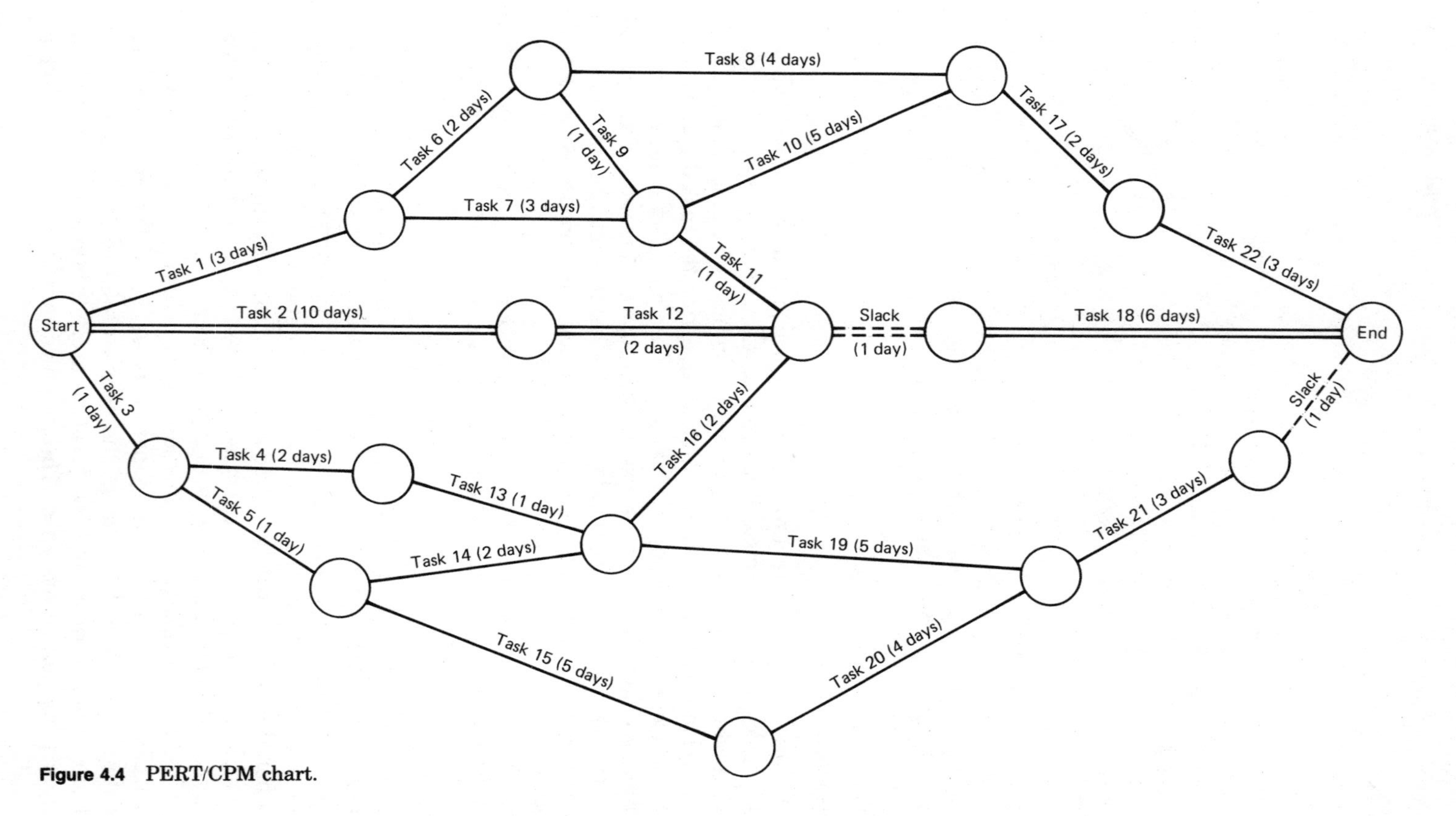

Figure 4.4 PERT/CPM chart.

Prepared by______________
Date prepared______________
Date revised______________

Project
Title____

Task number	Task description	Task estimated duration	Task skill level	Task assigned to	Period 1	Period 2	Period 3	Period 4	Period 5	Period 6	Period 7	Period 8	Period 9

Figure 4.5 Gantt chart.

Project Title____________________

Prepared by______________

Date prepared______________

Date revised______________

Task number	Task description	Task estimated duration	Task skill level	Task assigned to	Period 1	Period 2	Period 3	Period 4	Period 5	Period 6	Period 7	Period 8	Period 9
1	Task 1	3		1									
2	Task 2	10		2									
3	Task 3	1		3									
4	Task 4	2		3									
5	Task 5	1		4									
6	Task 6	2		1									
7	Task 7	3		5									

Figure 4.6 Gantt chart (tasks assigned).

Project Title____________________________

Prepared by______________
Date prepared______________
Date revised______________

Task number	Task description	Task estimated duration	Task skill level	Task assigned to	Period 1	Period 2	Period 3	Period 4	Period 5	Period 6	Period 7	Period 8	Period 9
1	Task 1	3		1									
2	Task 2	10		2									
3	Task 3	1		3									
4	Task 4	2		3									
5	Task 5	1		4									
6	Task 6	2		1									
7	Task 7	3		3									
1	John M												
2	Mary F												
3	George D												
4	Jane R												
5	Mark G												

Figure 4.7 Gantt chart (tasks assigned with personnel loading).

tation as in the upper chart, except here the period column contains a bar only when the upper chart indicates that the person is working on a task (Figure 4.7). The value of this lower chart is evident when it shows slack time for the project personnel, that is, when they are not working on any project.

Chapter

5

The Interview

CHAPTER SYNOPSIS

The interview is the primary technique for information gathering during the systems analysis phases of a development project. It is a skill which must be mastered by every analyst. The interviewing skills of the analyst determine what information is gathered, and the quality and depth of that information. Interviewing, observation, and research are the primary tools of the analyst.

This chapter provides a discussion of the interview and its importance, interview guidelines, and guidelines on interview documentation.

What Is an Interview?

A definition

An *interview* is "a formal face-to-face meeting, especially, one arranged for the assessment of the qualifications of an applicant, as for employment or admission. . . . A conversation, as one conducted by a reporter, in which facts, or statements are elicited from another." (*The American Heritage Dictionary*, Second College Edition)

The interview is the primary technique for information gathering during the systems analysis phases of a development project. It is a skill which must be mastered by every analyst. The interviewing skills of the analyst determine what information is gathered, and the quality and depth of that information. Interviewing, observation, and research are the primary tools of the analyst.

The interview is a specific form of meeting or conference, and is usually limited to two persons, the interviewer and the interviewee. In special circumstances there may be more than one interviewer or more than one interviewee in attendance. In these cases there should still be one primary interviewer and one primary interviewee.

Types of Interviews

During the analysis process, interviews are conducted for a variety of purposes and with a variety of goals in mind. An interview can be conducted at various times within the process for

1. Initial introduction
2. Familiarization or background
3. Fact gathering
4. Verification of information gathered elsewhere
5. Confirmation of information gathered from the interviewee
6. Follow-up, amplification, and clarification

Interviewing Components

The interview process itself consists of a number of parts.

1. Selection of the interviewee and scheduling time for the interview
2. Preparation of interview questions, or script
3. The interview itself
4. Documentation of the facts and information gathered during the interview
5. Review of the interview write up with the interviewee
6. Correction of the write up, sign-off, and filing

What Are the Goals of the Interview?

At each level, each phase, and with each interviewee, an interview may be conducted to

1. Gather information on the company
2. Gather information on the function

3. Gather information on processes or activities
4. Uncover problems
5. Conduct a needs determination
6. Verification of previously gathered facts
7. Gather opinions or viewpoints
8. Provide information
9. Obtain leads for further interviews

Interviewing Guidelines

Given these various phases and the variety of goals of an interview, the importance of a properly conducted interview should be self-evident. Since each interview is in fact a personal exchange of information between two personalities, a set of guidelines for the interviewer should be established to ensure that nothing interferes with the stated goal, i.e., gathering complete, accurate information. The interview is not an adversary relationship; instead it should be a conversation. Above all it is a process, and like most processes it has certain rules and guidelines which should be followed.

1. First and foremost, establish the tone of the interview.
2. Let the interviewee know the reason for the interview and why he or she was selected to be interviewed.
3. Stress that the interviewee's knowledge and opinions are important, and will aid in the analysis process.
4. Gain the interviewee's trust and cooperation early on, and maintain it throughout.
5. Establish what will happen to the information gathered.
6. Determine any areas of confidentiality or restricted information.
7. Let the interviewee know that candor and honesty will be valued and that nothing will be published or passed on until it has been reviewed and verified by the interviewee.
8. Firmly establish that there are no negative consequences to being interviewed.

Dos and Don'ts of Interviewing

The rules of interviewing are similar to the rules which govern most human interactions and to the rules which govern most investigative and problem-solving processes. In effect they can be called the rules of the game.

1. Do *not* assume anything.
2. Do *not* form prejudgments.
3. Do ask questions which start with who, what, where, when, why, and how, where possible.
4. Do ask both open and closed questions.
5. Do verify understanding through probing and confirming questions.
6. Do avoid confrontation.
7. Do act in a friendly but professional manner.
8. *Don't* interrupt.
9. Do listen actively.
10. Do take notes, but *don't* be obtrusive about it.
11. Do let the interviewee do most of the talking
12. Do establish rapport early and maintain it.
13. Do maintain control over the subject matter.
14. *Don't* go off on tangents.
15. Do establish a time frame for the interview and stick to it.
16. Do conclude positively.
17. Do allow for follow-up or clarification interviews later on.
18. Do be polite and courteous.

Who to Interview

One of the analyst's first and most important tasks during the data gathering phase of the analysis process, is to determine who has to be interviewed. This includes selecting the interviewees, understanding what can be expected from an interview of a person at a specific level, how to verify the information received from an interview, and, most important, understanding the perspective of the person being interviewed.

Most analysis projects have a user liaison assigned to the analysis team. It is this person's function to introduce the analyst to those being interviewed, to provide background information, and to interpret (or translate) the information which is obtained from the interviews. This person usually has the additional role of assisting the analyst in choosing those to be interviewed, scheduling the interviews themselves, and in some cases attending the interviews.

Under normal conditions, the analyst will have access to all people in the user area, although normally there is no need to interview them all. This is especially true if the user area is very large.

Generally speaking, the list of those to be interviewed can be divided into three sections: (1) the most senior manager, (2) his or her subordinates and junior managers, and (3) line workers, clerks, production people, sales staff, etc. See Figure 5.1 for interviewing and verification sequences.

The following are some guidelines for the analyst as to who to interview, when to interview them, what their perspective is, and what to expect from the interview (the goal of the interview itself).

1. *The most senior manager in the user area.* It is vital to interview this person at the start of the project. From him or her the analyst will obtain an overview of the user area, an overview of the functions performed by that area, and an idea of how the area fits within the overall structure of the organization and its activities.

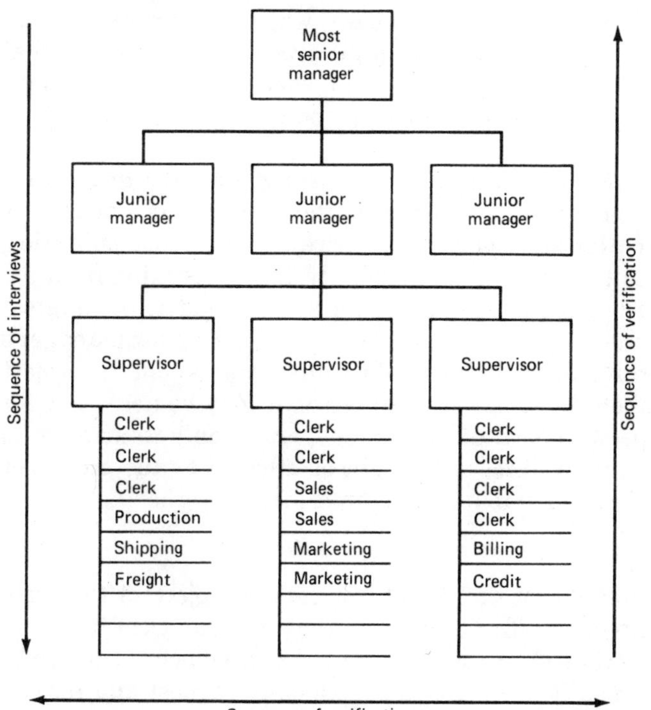

Figure 5.1 Interviewing and verification sequences.

The analyst can expect that the information received here will be general in nature. However, the manager will be able to define fairly accurately the business objectives of the project, the functions for which support is needed, any perceived problems which require attention, the time frame within which the project is expected to be completed, and the constraints, both business and financial, which apply. This manager can also suggest persons to interview, areas to concentrate on, and other sources of information. This manager should also be able to provide the enterprise perspective which is vital to the analyst's understanding

For each area within his or her control, the manager should be able to give the analyst an organization chart which indicates the structure of the area, the number of people within it, and an overview of its function. These charts should be used to select the names of individuals to be interviewed: the area manager, interviewees recommended by the manager, and, if necessary, alternative contacts in each area.

The analyst should not expect great levels of detail about the individual activities of the area nor about the individual tasks which are performed, much less how they are performed.

Regardless of any other information acquired, it is this senior manager who will benefit most from the project, who in most cases is funding the project itself, and who will in all probability have the final sign-off on the completed work. It is vital, therefore, to have a clear understanding of what is expected by this person.

2. *Immediate subordinates of the most senior manager and junior managers.* There may be many levels of these managers, each lower one having a smaller span of control, more specific responsibilities, and less authority than the one immediately above. Working from the organization charts supplied by the senior area manager, the analyst should schedule meetings with either (a) each of the senior manager's immediate subordinates or (b) those subordinates for whom the project is being undertaken. In either case (a) or (b), it may be necessary for the analyst to speak to each of these managers, if only to arrange to speak to people in their areas who might be affected by the project or who might be able to contribute to the information gathering process of the analysis.

It may also be found that the actual scope of the project, or the problem, is larger than originally defined and that the sources of the problem are in areas other than those which are experiencing the symptoms. For this reason it is usually a good idea to request and receive explicit permission from the senior manager to interview each of the second-line managers.

The senior manager should also be asked to explain to them the scope of the project, its intent, and that it may be necessary to interview some of their subordinates as the project progresses even though they are not directly involved (initially) in the project.

In many cases, people move from function to function, or at least from activity to activity during their employment by the firm. Many of these people may have more knowledge about the area under analysis than the current incumbents.

In many cases the managers themselves will have been promoted through the ranks, and will retain, if not current information about the area activities, at least some of the historical perspective as to how and why the area performs some of its activities and how they have changed over the years. In addition they can provide additional detail as to any problems, the reasons for those problems, and suggested ways to resolve them.

The managers at the various levels may also be able to clarify the statements made by their immediate superiors. The analyst will find that each person has a different perspective on an area, a perspective which reflects the individual's responsibilities and authorities. Each individual will also be able to provide a more detailed perspective on the interactions between different areas.

One of the things the analyst should be doing constantly during these interviews is verifying and cross-checking the information received. This cross-checking should be done in an objective manner and should not violate any information given in confidence. The objective is not to place blame or point fingers but to ensure that the information is correct. Where discrepancies arise between the information received from different managers, it may be necessary to verify the original information or seek a third source.

3. *Operational and clerical personnel.* These people actually perform the detailed tasks of processing, manufacturing, data gathering, or reporting. It is their tasks that the automated systems are ultimately intended to perform or augment. For this reason it is mandatory that the analyst thoroughly understand not only what they are supposed to do, but what they actually do, how they do it, and why they do it. It can be expected that most interviews will occur at this level. They will also be the most detailed.

Generally, the analyst will want to interview at least one person from each specific task or activity area. In some cases it may be necessary to interview more than one person. Each person interviewed will in all probability refer to something received from or passed to another person or area. The analyst must verify that both individuals agree on the identity of the materials they are passing to each other and that both agree on the content of the transferred material.

The understanding of the operations level personnel as to their specific duties, functions, and activities must match that of their immediate managers. Any discrepancies in this area must be resolved by the analyst.

The Need for Documentation

Everyone talks about the weather but no one can do anything about it. In the case of documentation, everyone talks about it but few do it; however, unlike the weather, most people can document, and document effectively.

Documentation, however painful and tedious it may seem, is one of the most critical tasks of analysis. The documentation produced as a result of the analytical interviews, the analyst's observations and research, and ultimately, the total analysis phase of the project serves a number of purposes.

Permanence. The need for documentation is rooted in semantics and human memory. Verbal communications are both transitory and subject to interpretation. The average person has a language working set of about 500 to 1000 words. The written working set, by contrast, is much larger, perhaps by as much as an order of magnitude. Verbal communication is augmented by inflection, body language, and by a process of feedback and interaction, all of which serve to clarify the ambiguous, the ill-defined, and ill-understood. Human memory is imperfect. Words communicated verbally can only be recalled and examined with difficulty, if at all.

Precision and recall. A written document is more precise and may be reviewed repeatedly until understanding is achieved. It has the added advantage that small changes can be made to it without having to restate the entire premise or thought. Additionally, once an idea is written down, it may be recalled at will exactly as first presented and may be completed by someone other than the original author, or authors. Because there is little feedback from the written word, one can only take issue with misstatements of fact or with ambiguous wording. If it isn't written down, it isn't there.

Graphics. Documentation usually includes both a narrative portion and an illustration portion. These illustrations serve to amplify and enhance the narrative. One picture can be worth many thousands of words, if properly drawn. The graphics of the analytical documentation, whether it employs simple flowcharts, Hipo diagrams, data flow diagrams, or powerful modeling techniques such as those based upon the entity-relationship-attribute approach, presents the user and the analyst with a way to walk through the picture developed from the analysis, and ultimately walk through the design developed from the analysis. These walk-throughs enable both to understand the environment and to detect any ambiguities and anomalies. Illustrations have the added advantage that they can be viewed in their entirety, whereas narrative may only be viewed in fragments.

Functions of Documentation

Documentation serves to clarify understanding, and perhaps most important, it provides the audit trail of the analyst. That is, it creates the records which can be referred to at some later date and which serve as the basis for future work and decisions.

Good documentation precludes the need to return to the interviewee for a repetition of ground previously covered. Good documentation can be reviewed over and over until adequate understanding is achieved.

Documentation is tedious and sometimes boring. But it is also vital. Good documentation allows other analysts and the analyst's successors to pick up where the first left off, should he or she be reassigned. Documentation is necessary if the next project phases are to be successful, since they are predicated on the results of the analysis. To a very real extent, analytical documentation provides the road maps for the remainder of the project. If the maps are faulty, or incomplete, the succeeding teams may wind up in a swamp, or worse, in quicksand.

Most important, the finalized documentation serves as a contract between the user and the data processing developer. In it the analyst has described the user's environment, the analyst's understanding of the user's needs and requirements, and with the proposal for a future environment, the analyst's description of the system to be designed and built by the developers. With the user's sign-off, or approval of these documents, a contract is created between the two. Barring unforeseen changes in the business environment, the problems described in the documentation will be rectified and the environment proposed will be the one built and installed for the user.

The document becomes, in effect, a statement of the work to be performed. The time to modify and change it is before the work begins; afterward it may be too late. From the developer's perspective, any post sign-off changes may require a renegotiation of either time frames, costs, or resources. From the user's perspective, the design is what is contracted for and what he or she is paying for. If the final product does not conform to the proposal, then it is up to the developer to rework the product until the user is satisfied.

Documenting the Interview

The interview is not complete until it has been documented. See Figure 5.2 for a header form for sample interview notes. The documentation of the interview need not be a verbatim transcript of what was said but should cover the following items.

1. Who was interviewed and who did the interviewing, including the title of the interviewee, the interviewee's function and immediate superior, and interviewer name and title

Interviewee ____________________

Title of interviewee ____________________

Immediate supervisor ____________________

Interviewer ____________________

Interviewer title ____________________

Others present ____________________

____

____

____

____

Date ____________________

Time ____________________

Location of interview ____________________

Interview ______ of ______

Interview objective ____________________

____

____

____

Interviewee's supervisor

Interviewee and peers

Interviewee's subordinates

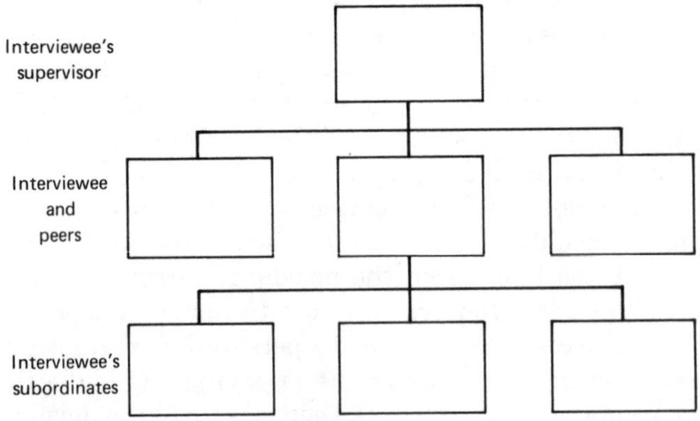

Figure 5.2 Sample interview notes header form.

2. The date, time, and location where the interview took place
3. Names and titles of any other persons who were in attendance
4. The stated objective of the interview
5. The interviewee's job or functional description
6. The interviewee's organizational chart and organizational charter, if appropriate. A complete description of any facts which were obtained during the interview
7. A complete description of any opinions stated by those interviewed
8. Any conclusions drawn from the facts or opinions as presented
9. Any business problems uncovered during the interview
10. Samples of any forms, reports, charts, graphs, documents, manuals, policies, standards, or procedures referred to or discussed during the interview
11. Any charts or diagrams drawn as a result of the information gathered during the interview
12. Any relevant comments made by the interviewee
13. Any numbers, transaction or document volume information, task timing information, capacity information, quality information, etc., gathered during the interview

Dos and Don'ts of Documentation

1. Keep a clean separation between fact and opinion, whether the opinions are those of the user or those of the analyst. In your documentation, be sure to identify clearly which parts are fact and which are opinion.
2. Document all sources of information, especially all facts and numbers.
3. Make sure that all documentation is dated, not only with the date of the original version, but also with the dates of any revisions. If documentation has been revised, a revision identification scheme should identify which revision is most current. See Figure 5.3.
4. Ensure that all pages of the documentation are numbered and that there is a table of contents which indicates where each major section may be located.
5. Analytical documentation tends to be long, sometimes overly long; however, it is better to err on the side of length than to leave out information which may become important later on.

Project name ______	Date prepared ______
User sponsor ______	Date revised ______
Author ______	Section no. ______
Source ______	Page ______ of ______
Verified by ______	Revision no. ______
Subject ______	

Figure 5.3 Typical documentation page header information.

6. Write for your audience, which implies that your audience should be known. Technical information should be written for technical audiences; user documentation should be written for users.
7. All documentation should contain an executive summary which contains salient background material, findings, conclusions, and recommendations, in that order. This executive summary material should be explained in depth in the support materials. See Figure 5.4 for a major section table of contents as it would appear in the documentation.
8. Studies have shown that most people will only be interested in the first 20 to 30 pages of any report. All the information you need to convey should be in those first 20 to 30 pages; any back-up material should be reserved for appendices and supporting documentation.
9. One picture is worth a thousand words. Pictures could include graphs, charts, tables, and diagrams. They are highly informative and easily understood. Use them liberally. The documentation should contain a listing of the location of all graphs, charts, and

Project name		Date prepared	
User sponsor		Date revised	
Author		Section no.	
Source		Page ____ of	
Verified by		Revision no.	
Subject	Table of contents		

Table of contents

I Table of contents
II Executive summary
III Findings
IV Conclusions
V Recommendations
VI Cost benefit analysis
VII Tables and illustrations
VIII Glossary
IX Index

Figure 5.4 Documentation major section table of contents.

diagrams. *Remember:* When using charts, diagrams, etc., that all special symbols should be identified in a legend, along with the source of the information for the charts, diagrams, etc. All axes of charts and graphs should be clearly labeled. Chart and graph scaling information should be clearly in evidence. All pictures should be clearly titled.

10. Acronyms may be used, but they should be spelled out in full at their first use. The documentation should contain a list of these acronyms along with their full word meanings. If in doubt, spell it out.
11. All businesses and functions have their special terminology. Its use is unavoidable, and in fact, it would be unwise not to use it. All documentation should contain a glossary which defines these special terms.
12. The best analysis can be discredited by sloppy documentation. The documentation doesn't have to be professionally done, but it should be neatly typed, with no spelling errors or typos. The

grammar should be reasonably correct. The final document should be neatly bound with cover pages.

13. Avoid verbosity. Avoid pomposity. The document should be written in business English with reasonably short sentences. This is not the place to try and impress people with the extent of your vocabulary. Use headings and subheadings liberally.
14. Most large documents will have sections which are written by more than one person. Try to ensure some level of stylistic consistency. The document should flow and should be easily readable. Remember it is meant to convey information.

Report Format Style Guidelines

The appearance and readability of documentation can be greatly enhanced by following these guidelines.

Use a consistent outline format. For example,

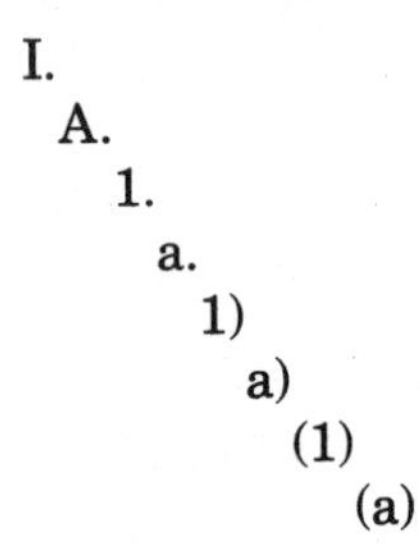

If it becomes necessary to itemize under any heading use lower case roman numerals.

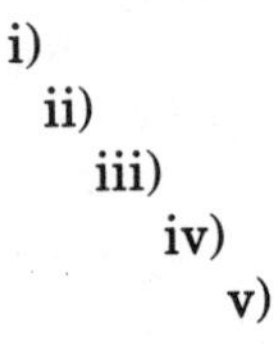

An indentation should always have more than one item.

Every outlined section should always have a boldfaced heading.

Do not start a paragraph with a number or a single letter.

Avoid the use of bullets or special marks in final reports—they give the impression that the report is still a summary

Make sure that all words are properly spelled and that proper grammar is used. Nothing can be more detrimental to a report than improper grammar or word usage. Avoid slang, technical jargon, and idiom where possible.

The following are general stylistic considerations which apply.

1. Never use the term "etc." or "et al." in a report. If the other items are known list them. If the list is too long to include in the text, or if the items are relatively minor, put them in an appendix.
2. Do not use the terms "same as" or "ditto" in the final text. Always use sentences and paragraphs, and avoid freestanding phrases.
3. All exhibits should be meaningfully titled and referred to in the text. Exhibits should follow, or be as close as possible to, the page of reference.
4. All financial numbers should be checked for arithmetic. That is they should cross and down-foot.
5. All exhibits should be consistent from exhibit to exhibit, from exhibit to text, and chapter to chapter. Be sure to use consistent units of measure and time frames in all exhibits.
6. When numbers are used within the text, make sure that the units of measure are indicated clearly. Single number references should be spelled out in words and followed by their numeric equivalent, e.g., six (6), thirteen (13), twenty-two (22). All monetary figures should be preceded by the appropriate currency symbol and show all commas and decimal points for clarity (i.e., $1,000.00)

Chapter

6

The Data Dictionary

CHAPTER SYNOPSIS

An automated dictionary is one of the most powerful tools available to analysts in terms of documenting the information gathered from their activities. The dictionary can hold these analytical findings and can also be used to develop cross-references and correlations between data and information items.

This chapter discusses the functions of a dictionary, how a dictionary is constructed, and the roles of the database administrator and the data administrator in the analytical process.

Introduction

One of the primary products of the analytical process is documentation. Some large projects produce large amounts of documentation, some of it running into hundreds of pages. Although development methodologies provide some structure to the documentation, even the best methodology and the most meticulous analyst cannot organize, index, and cross-index the information sufficiently to make it entirely useful. There is always the need to see the information differently from the way in which it is presented. Even with the best organization and indexing methods, hardcopy documentation (documentation printed on paper) is difficult to revise and still more difficult to keep up to date.

In order to overcome these and other purely mechanical problems involved with producing and maintaining proper documentation, many firms rely on automated dictionaries. These data processing system products are specifically designed to hold, maintain, and organize analytical information; they come equipped with flexible facilities for producing a wide variety of reports on the dictionary contents.

Why a Dictionary?

Although automated dictionaries are usually discussed within the context of database administration, data administration, or database in general, it is equally appropriate to discuss them in terms of the systems analysis task process. The reasons for this become obvious when one looks at what they are designed to do and what they are designed to contain.

These dictionaries, usually called data dictionaries or data dictionary/directories, are in reality application systems which have been designed to accumulate documentation. Although the majority of the documentation within them pertains to data, most of them also contain provisions for documenting the other components of the systems environment: systems, users, reports, forms, functions, processes, etc.

Most people do not look at them as systems but rather as automated tools, usually implemented under a DBMS, which are designed to contain system documentation and to facilitate management of data. Dictionary systems and their files, unlike most systems, do not contain data, but rather they contain "data about data." This "data about data," usually called "metadata," describes and defines the components of the data processing environment and allows for each type of component to be related to every other type of component, thus creating a powerful research, analytical, and cross-reference tool.

Because of their automated nature, and normal ease of updating and maintenance, dictionaries are the primary documentation tool for most well-run data processing organizations. In some cases the dictionary is also the repository for the control and definitional information which drives the DBMS. In these cases the DBMS software is constantly interrogating the dictionary for information. Because any changes made to the dictionary are immediately reflected in the operations of the DBMS , and because no change can be made to the DBMS environment except through the dictionary, these "active" dictionaries are an integral part of the operational software of the DBMS product and help to ensure consistency of data definition and usage.

The opposite of an "active" dictionary is a "passive" dictionary. These products provide the same documentation capabilities as their active counterparts but are completely separated from the DBMS product. Because of these differences between the two types of dictionary products, active dictionaries are normally provided by the DBMS vendor, whereas passive dictionaries may be provided by the DBMS vendor or by an independent product vendor.

What Is Data Administration?

Within most companies where a DBMS product is installed, there is a functional unit or organization which is usually part of the database support environment and which is assigned responsibility for updating and maintaining the dictionary. These organizations usually have the additional responsibility for the functions of "data analysis" and "logical database design."

The name given to this part of the organization is data administration. Generically, data administration is that organization which is assigned responsibility for gathering information about corporate data; for verifying the definition, source, and usage of that data; and for preserving that information for the company and making it available in a timely way to all company personnel who may need it.

Normally data administration is *not* responsible for the capture, verification, processing, storage, retrieval, or dissemination of company data, only for the data about data, the definitions, descriptions, and so on. Within the data administration purview would also be the responsibility for determining firmwide consensus on coding structures, naming conventions, or naming standards.

Data administrators work with analysts to define both the data elements required for the business function and the structure of that data. Data administrators assist analysts in the process of data analysis and in documenting that analysis into the dictionary.

Since all information processing systems require the acquisition and manipulation of data, most analysis is devoted to identifying that data, its sources and uses, and the processes applied to it. Because of the critical role data plays in most firms, data administration has also become very important, and data administrators routinely play an important role in assisting the analysts with their documentation. In effect data administration has become the documentation center of most firms. In many cases the data administration function has also been assigned the responsibility for developing the system methodologies and for ensuring that they are followed.

How Is a Dictionary Designed?

Because most firms have some form of automated dictionary, it is appropriate to discuss how they are designed and thus what they are capable of containing. Most data dictionaries are composed of a number of related files, sometimes in database form, which store the documentation for application systems and their definitional components. A conceptual model of automated dictionary documentation categories and relationships is shown in Figure 6.1. Data dictionary files are usually designed such that each file or portion of a file contains data about some aspect of the systems environment. Each file is thus tailored to document a specific system component. These components form the bulk of the firm's documentation needs for its automated systems. Although most dictionary systems are capable of documenting each major component, not all firms insist that each part be used.

Data elements

The first and most detailed piece of documentation within the dictionary is the data element. Data elements, sometimes called fields, are the lowest unit of meaningful information within the business, and by extension, within the data processing environment (Figure 6.2). Data elements have characteristics called attributes: size, shape, format, value range, content validity, and location. In addition data elements have meaning from a business sense. Data elements can be natural, coded, or derived.

1. Natural elements contain data which is unchanged from its natural usage, such as name, address, social security number, or telephone number.

2. Coded data elements come in a variety of forms ranging from normal business abbreviations, such as NY to mean New York, or F to mean female, to firm-specific codes such as the numeral 1 to mean cash account. When coded elements are documented, that documentation must be accompanied by a key which is used to decode the valid code values into their natural form.

When dealing with coded elements the analyst must first identify which elements are coded, then determine if the complete code list exists, verify that list, and determine if the coding structure needs to be expanded.

3. Derived data elements are those where the field values are determined as a result of some formula, such as,

Net profit = gross sales − (expense of sales + overhead expense)

Here the analyst must determine which data elements are derived, document the formula for derivation, document the functions of that formula, and determine if the formula needs to be modified or if it can be simplified or rederived.

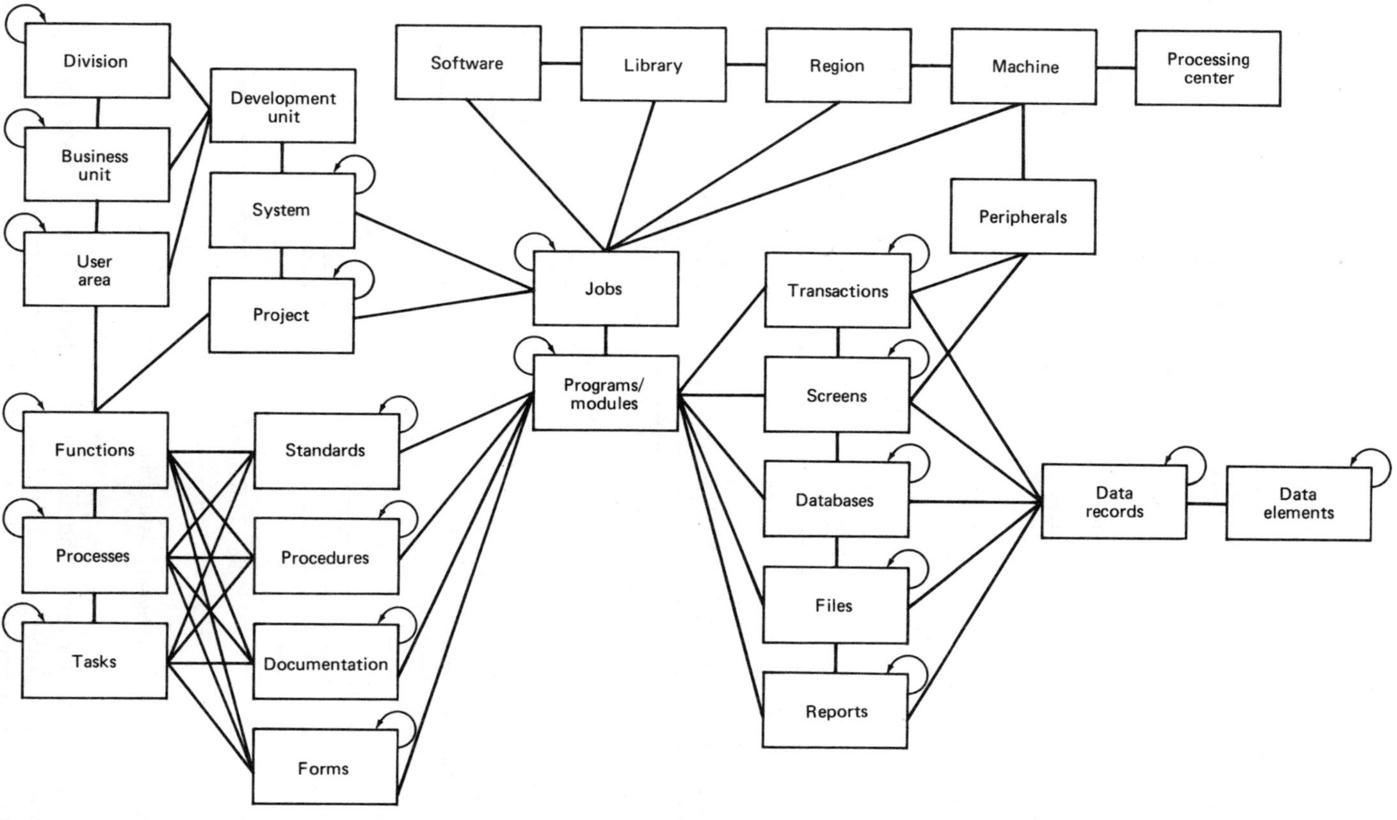

Figure 6.1 Conceptual model of automated dictionary documentation categories and relationships.

BUSINESS DATA ELEMENT DESCRIPTION WORKSHEET

Prepared by ______________________
Project ______________________
Date prepared ______________________

Dictionary Action:
ADD——— CHG——— DEL———

Global Business Name (72 characters maximum)

COBOL Data Element Name (26 characters maximum)

Derived (Y or N) ———

Business Data Element Description:

Standard Values (if applicable):

CODE	DESCRIPTION	CODE	DESCRIPTION

Formula for the Derivation of This Business Data Element (if applicable)

Business Data Element = ______________________

where: ______________________
E1 = ______________________
E2 = ______________________
E3 = ______________________
E4 = ______________________
E5 = ______________________

For Data Administration Use Only

Processed by | / / Date

Figure 6.2 Sample dictionary data element description work sheet.

Part of the power of the dictionary is derived from its ability to relate various pieces of documentation to other pieces of documentation. Data elements are related within the dictionary to

1. Records which contain those elements
2. Files which contain the records which contain the elements

3. Programs which use those data elements
4. Systems which use those data elements
5. Users who own, maintain, or use the data elements
6. Other data elements

Records

If the data element is the most detailed documentation item within the dictionary, then the next most detailed items are records. Data records come in a variety of forms. In essence any aggregate of data elements can be considered a record. Data records are usually a single entry in a file of like entries. They can also be a single instance of a report page, a single folder in a file of like folders, a single form in a file of like forms.

Data records may contain as few as one data element, and there is no practical upper limit in terms of maximum data elements. Documentation of records consists of descriptions of the record, its identification by name, type, use, source, distribution, frequency of creation, etc. As with data elements, within the dictionary records are related to

1. Data elements which are contained within those records
2. Files which contain the records which contain the elements
3. Programs which use those records
4. Systems which use those records
5. Users who own, maintain, or use the data within the records
6. Other records which are related to the record under study in some way

Data files

Just as many different types of data elements are aggregated to records and forms, so too, many different types of data records and forms are aggregated to data files. Data files are in fact collections of like or similar records. Data files normally contain all occurrences of data records on a like subject, such as all employees, all payroll records, all accounts, all securities, etc.

Data files may contain as few as one record, and there is no practical upper limit in terms of maximum records. These ranges apply both to the types of records and to the numbers of occurrences of each type. Data files, like their component records, are named, described, and identified as to source, use, ownership, frequency of use, storage medium, location, etc. Data files are related within the dictionary to

1. Data elements which are contained within the records which constitute the file
2. Records which are contained within the file
3. Programs which use the file
4. Systems which use those files
5. Users who own, maintain, or use the data within the files
6. Other files which are related in some way to the file that is under study

Procedures

Data elements, data records, and data files do not exist in a vacuum. They must be associated with procedures which define how they are to be used, who uses them, how they are to be processed, when they are processed, where the data comes from and where it goes, how long it is to be retained, how it is to be verified and validated, and how to handle exceptions to normal processing rules.

A procedure then is a formalized method of accomplishing a given task or set of tasks. It sets down, in a step-by-step fashion, that which must be accomplished, when it must be accomplished, how it must be accomplished, and by whom it must be accomplished. Procedures detail methods of handling inputs and generating outputs, or simply methods for analyzing file and record contents and making decisions based upon that analysis.

Automated procedures, programs, and program fragments (called modules) provide the detailed step-by-step instructions which direct the computer's central processors to take the appropriate actions on the input data, and to manipulate that data to produce the desired outputs. Procedures, both manual and automated, contain the rules and guidelines for accomplishing a given task or set of tasks. Procedures are related within the dictionary to

1. Data elements which are input to or output from the procedure
2. Records which are processed by the procedure
3. Programs which are used by the procedure
4. Systems which are associated with the procedure
5. Users who are responsible for accomplishing the procedure or who maintain the procedure

6. Other procedures which are related in some way to the procedure under study

System

A system is a collection of procedures, automated, manual, or both, which has been assembled and organized to accomplish the tasks, activities, and processes associated with a user function or subfunction.

Systems are documented as to what processes, procedures and activities are encompassed by them, who is responsible for their performance, who initiates them, who supplies the inputs, who receives the outputs, what files are associated with them, what their purpose is, and what functional areas they support. Systems are related within the dictionary to

1. Data elements which are input to or output from the system
2. Records which are processed by the system
3. Programs which are used by the system
4. Procedures which are associated with the system
5. Users who are responsible for the operation of the system or who maintain the system
6. Other systems which are related in some way to the system under study

User

A user is usually someone outside the data processing area for whom a project is undertaken. Users have primary corporate responsibility for the accomplishment of functions, processes, and business activities. Users supply analysts with the background, and general and detailed knowledge about the tasks which need to be performed and about the business problems which need to be solved.

Users are documented in terms of their functional responsibility, the systems that service them, their reporting structure (to other users), the files which they are responsible for, the transactions they generate or process, and the reports they generate or use. Users are related within the dictionary to

1. Data elements which are required for any task which they perform
2. Records which are processed by them
3. Programs which are used by them

4. Procedures which are associated with them or which they use in their day-to-day work
5. Systems which they operate or which are operated for them in order to accomplish their tasks
6. Other users which are related in some way to the user in question

Other dictionary contents

Dictionaries may also be used to document user functions, processes, activities, and tasks independently of the user areas themselves. Other items and categories of information within these automated dictionaries can be the organizational units and structure of the firm itself, and the physical resources of the firm as they relate to information processing. In some cases the dictionaries have been used to contain the standards, guidelines, and procedures which govern the development process itself, as well as information relating to the types and locations of documentation not contained in the dictionary.

How Does a Dictionary Function?

Automated dictionaries are fundamentally application systems. In this case they are applications which assist the development unit itself. As with any other system, especially purchased ones, they may have to be modified by the specific firm for its own use.

A dictionary consists of a collection of files designed to store dictionary data and a collection of programs, batch, online, or both, which perform the tasks of data input, data manipulation, and data presentation. A dictionary may have programs and procedures associated with it which were written for the specific installation.

Generally speaking the process of dictionary use begins with the gathering of information about some aspect of data or other category of information within the dictionary. This is written down, verified, and codified to suit the particular dictionary being used. (See Figure 6.2 for an example of a dictionary element data collection form.)

After the data are gathered, they are either given to the data administrator for entry or may be entered by the analyst or the user, assuming he or she has sufficient training in dictionary use.

Once the primary entry is in the dictionary the analyst or data administrator begins the process of editing the information and associating that entry with every other known entry to which the new entry is related. As new entries or new information about old entries are gathered, they are also entered into the dictionary and related in a similar manner.

How Can the Dictionary Assist the Analysis Function?

Progress reporting and documentation

The data administration staff produces periodic reports for the analyst listing all entries associated with the analyst or the user to date and their cross reference information. These reports can be produced in any order, and from any vantage point within the dictionary configuration. That is by element, by record, by file, by procedure, by system, by user, or by any attribute of any of the above. These reports may be all-inclusive or may be restricted to some subset of the data entries.

Because of the wide variety of entries which can be stored in the dictionary and because of its ability to store narrative definition on each entry, the dictionary can act as a primary documentation tool for the analyst. The analyst may also use the online capabilities of the dictionary to check on any reference.

Research

Because the dictionary contains not only data entered by the analyst doing the research but all data entered by other analysts, and about all systems, the analyst can use either the online or batch-reporting capabilities of the dictionary to browse its contents. Using this capability the analyst can determine if any of the entries about to be made are already entered, in which case only the additional information need be added.

In some cases the analyst may be searching for some particular data to service a user, and the dictionary can be used to determine if it is already being captured, and if so, by whom, and when, and where.

Conversely if the analyst is seeking to determine the impact of a proposed change, the dictionary can be used to find all known users and uses of the item to be changed.

Chapter

7

The Entity-Relationship Approach

CHAPTER SYNOPSIS

The entity-relationship approach to analysis is a relatively new technique—one which is not widely understood or used. By focusing on the business and data entities of the firm, their relationships to each other, and the attributes which are needed to describe the entities and qualify their relationships, the analyst focuses on the real world, people, places, and things of the business—things and relationships which can be readily observed and documented.

This chapter discusses the theory and concepts behind the entity-relationship approach, and how it can aid in focusing the analysis and in identifying entities, relationships, and attributes. The various levels of entity-relationship analysis are also discussed.

The Entity-Relationship Approach

The entity-relationship approach consists of an analytical method and a modeling technique. Although the entity-relationship (ER) approach was first described by Dr. Peter Chen in a paper which appeared in the first issue of the ACM publication, *Transactions on Database Systems*, in 1976, it has only recently been recognized as one of the most important tools in the data administration tool kit. Properly understood and used, the ER approach can greatly reduce the time needed during the analysis phase of the development cycle and at the same time greatly increase both the accuracy and completeness of that analysis.

Most popular analytical approaches, or methodologies, focus either on the processes being performed or on data elements presumed to be needed by the user. Some concentrate on trying to fit lists of data elements into one of the data structure models which can be implemented by a DBMS, others on designing reports, screens, and files, and still others on following trails of transactions through the various processing stages. From these processes, flows, data elements, and/or outputs, they attempt to re-create the real world. Many attempt to re-create the processes from the desired results.

The major flaw of these methods is that their focus is too narrow. They assume that because we are data processors, the world is only made up of data. Each method approaches the data problem differently, and their results, many times, resemble the results of the blind men who examined the elephant: the one examining the tail thought it felt like a rope, the one examining the sides thought it felt like a wall, the one examining its legs thought it felt like a tree, and the one examining its trunk thought it felt like a snake. In a sense they were all right, but, at the same time, none was right.

In the business environment, examining only transactions, processes, outputs, or data flows, or even a combination of all four, produces a picture which is correct as far as it goes, but which is not a true or complete picture of the environment. The business environment is also populated by people using things, and both people and things are located in places. Any business description must not only include these people, places, and things, but it must also start with them. These people, places, and things are called entities. These entities interact with each other in various ways, and those interactions are called relationships.

People, either individually or in groups, work with things or provide services for other people. Since both the people and the things are real (they physically exist), they must be described and they must be located somewhere (in some place). Additionally, relationships which exist between people and things, people and places, things and places, different types of things, different types of places, and different types of people themselves must be described.

These entities may be well defined, in that the firm may know a great deal about them, or they may be vaguely defined, in that the firm may know very little about them. In some cases, such as with either prospective customers or employees, the firm may only know or suspect that they exist, but not who or where they are.

These entities may exist in large homogeneous groups where all members are capable of being described in the same manner, or they may be fragmented into many different subtypes, each with descriptions which are either slightly different, or in some cases radically different, from other members of the same group.

Just as entities are real so are the relationships that exist between them. And as with entities, the relationships may be well defined in that the firm may know a great deal about them, or vaguely defined in that the firm may know very little about them, as little as first knowing or suspecting that they exist.

The power of the ER approach lies in its ability to focus on describing entities of the real world of the business and the relationships between them. By describing real world entities through the identification and assignment of attributes to them and their relationships, the analyst is describing how and why the business operates.

Although the business itself may change, sometimes dramatically, these types of changes occur much less frequently than changes in the routine processes and activities. Regardless of how the business changes, the entities of the business rarely change. What may change, however, is the firm's perception of which attributes of those entities are currently of interest. Some relationships between these business entities may also change, but even these relationship changes occur infrequently. Thus by understanding and properly describing these entities and the relationships between them, the analyst can form a very stable foundation for understanding and analyzing the business itself, and for properly recording the results of, or changes caused by, the processes of the business.

As with any analytical method, the effectiveness of the ER approach is limited, or constrained, by three factors, all of which have to do with the analyst's understanding of the business environment. These closely related factors are (1) entity identification, (2) entity definition, and (3) business context.

Entity identification consists of recognizing the various entities, determining why they are of interest to the firm, and naming them. The identification process must specify the entity at the precise level of specificity which ensures that it is not so general as to be meaningless, and yet not so specific that it fragments into too many subsets. For example, people as an entity would be too general since it includes both customers and employees, among others. On the other hand, full-time employees and part-time employees would be too specific since both are employees and "full-time" and "part-time" are attributes of employee.

Entity definition consists of identifying which attributes of the identified entities are needed by the firm and why those attributes are of interest. For example, is the firm interested in the attribute "hobbies" or "clothing sizes" for the employees? If the firm deals in sporting goods, the answer to the former might be yes. If the firm provides uniforms for its employees, the answer to the latter might also be yes.

Business context involves identifying and defining the relationships which exist between the identified and defined entities, and their relative importance to the firm as a whole and to each of its specific parts. Business context also involves identifying and defining the use or role of each entity within the firm. An entity's appearance, role, or use in one firm may be entirely different in another firm; yet the entity itself is the same.

Just as an entity may have different roles or uses between firms, so also, each part of the firm may have a different perspective on the business, and, consequently a different perspective on the entities of the firm. This perspective does not change the fact of the entity's existence, only the attributes and relationships of those entities which are of interest to individual portions of the firm and their role or use in that firm.

The specific definitions of these entities and their relationships with other entities within the firm are relevant only within the context of that firm and are totally dependent upon the attributes of the entities which are of interest to the firm. An entity within one firm may be only an attribute of an entity within another firm, and vice versa.

The importance of identification, definition, and context can be seen when one looks at the formal definitions of the three key elements which form the heart of the ER approach. These definitions form the basis for both the data analysis method and the data modeling technique of the entity-relationship approach.

A definition

An *entity* is defined as a person, place, or thing which (a) is of interest to the corporation, (b) is capable of being described in real terms, and (c) is relevant within the context of the specific environment of the firm.

A definition

An *attribute* is any aspect, quality, characteristic, or descriptor of either an entity or a relationship. An attribute must also be (a) of interest to the corporation, (b) describable in real terms, and (c) relevant within the context of the specific environment of the firm.

An attribute must be definable in terms of words or numbers. That is, the attribute must have one or more data elements associated with it, one of which may be the name of the entity or relationship. An attribute may describe what the entity looks like, where it is located, how old it is, how much it weighs, etc. It may describe why a relationship exists, how long it has existed, how long it will exist, or under what conditions it exists.

A definition

A *relationship* is any association, linkage, or connection between the entities of interest to the corporation. These relationships must also be (a) of interest to the corporation, (b) describable in real terms, and (c) relevant within the context of the specific environment of the firm.

It is important to note at this point that relationships exist only between entities, not between attributes of entities. To illustrate,

> The entity "person" could be anyone.
>
> When the attributes name, age, and sex are added, we can distinguish men from women, adults from children, and one person from another.
>
> When the relationships are added, we know whether we are talking about a group of unrelated people, a family, or a corporation.

Entities and their attributes

To describe an entity, we must describe it in terms of its attributes and its relationships with other entities. An entity description consists of a series of statements which complete a phrase such as "the entity is . . . ," "the entity has . . . ," "the entity contains . . . ," or "the entity does . . . ."

Each attribute relates to the entity in hierarchic terms, that is, all attributes of the entity are fully dependent upon the entity itself because individually and together they are the entity. The question can still be asked, however, "How can we begin to identify these entities?" Is, for example, the entity identified as customer (representing all customers), or is it the specific type of customer (such as mail order or retail), or is it a single customer? The answer is that it can be all of these, none of these, or more than these.

The specific identification of the entity has meaning only within the context of that firm. However, most businesses can be described using a fairly restricted set of generic entity types such as customer, product, machine, employee, location, organizational unit, etc.

An entity is whatever the business defines it to be, and that definition must make sense within the context of the firm. Thus, an entity in one firm may be a subset of entities included in the entity definition of another firm, or may be the global definition of the entity used within another firm. These differences in identification can be illustrated by the following example.

> If we were a town planning board, with responsibility for community planning and zoning, we could describe that community in terms of its buildings, and further subdefine those buildings into residences, offices, stores, warehouses, and factories.
>
> We might be interested in which people or firms occupy or own those buildings, but for our purposes that information would be an attribute of the building, just as the size of the building, the number of floors, the number of windows and doors, and the cost of the building are attributes.
>
> On the other hand, if we were the local tax assessor doing a census or community directory, we would be interested in the people who live and work in the community and firms located there. In that case we would be interested in the names of the people, their incomes, length of residence, amount of taxes paid, and where they live or are located (the buildings) within the community. Here the buildings become attributes of the people.
>
> Neither the buildings nor the people have changed. They both still exist. Our perspective, however, has changed; the things which interest us about those buildings and people have changed.
>
> If we raise our perspective to that of the town council, then we need to know all the information about both the people and the buildings, along with information about roads, utilities, etc. In this case, both the buildings and the people become entities in their own right, along with the relationships between them (who lives or works where, who owns what, and so on).

This need for both attributes and relationships is consistent with the accepted dictionary definition of an *entity*: "the fact of existence; being. The existence of something considered apart from its properties." Thus although the entity exists, its true form and role are only apparent after its attributes are added.

Without attributes all we know about the entity is that it exists. The distinction between an entity and its attributes, and the relationship between an entity and its attributes, is so important that the ER diagram distinguishes between an entity and its attributes by using different symbols for each.

The entity-relationship model

The entity-relationship model represents a conceptual view of the world; as such it is independent of any DBMS or data processing considerations. It is a creature not of data processing but of the business environment.

Although we speak of an entity as if it were singular, in reality it is that set of persons, places, or things which have a common name, a common definition, and a common set of descriptors (properties or attributes).

The entity representation in the model, while it may represent a single instance, usually represents numerous people, places, or things which have a common name and common descriptors and thus can be treated as a set. These entities interact (relate) with other entities. These interactions form a complex set of relationships.

An entity, although it exists physically, only has physical substance when it is described in terms of what it looks like, where it is, what it does, and how it relates to other entities. Each component of that description is a property or an attribute of the entity. The sum of the properties is the entity.

These entities are physically real and their real properties can be described; these people perform actions, using and transforming both things and information (which is contained on things as data). The common characteristic of all entities is that we can describe them, and we use words and numbers for that description. Collectively these words and numbers are data; individually they are data elements.

The fact that entities, especially in the data processing environment, are described by data, does not make them data objects nor is every collection of data elements an entity.

Some writers have suggested that data entities are built from collections of data elements in the same manner that a car is built from a collection of parts. In fact, an ER model can be complete and meaningful with no traditional data elements at all. The parts of a car were specifically chosen because each contributes something to the overall design of the vehicle. Any number of different sets of parts could be assembled and would result in a car, but a specific car can only be built from a specific set of parts.

A car is a thing. It is a subtype of the larger group of things called vehicles, and part of another subtype called self-powered vehicles which transport people and things. Just as there are many different types of vehicles, not all of which are cars (some may be boats, planes, or trains), so too there are many different types of entities.

A final type of attribute needs to be discussed: attributes which do not describe the thing itself, but what it does, how it is used, or why it is used. Those things that an entity does are called activities; collectively they are called processes. The attributes which describe these entity activities are called processing-related attributes.

The processes, or activities, of the business are in reality the actions that people take with respect to things, places, or other people. These actions usually result in some change in the physical appearance, state, or condition of one or more entities, or sometimes in the creation of a new entity. We can use the entity called car as an example.

> The physical characteristics of the car, its size, weight, year, make and model, color, and parts list represent the car itself. Whatever happens to it, so long as it remains a car, these characteristics (except for possibly color) will never change. Whether it is owned by anyone or not, new or used, in good repair or falling apart, driven 1 mile or 100,000 miles does not change the fact of its existence. However, the fact that the car exists is meaningless unless we put it in a context which tells us why the firm is interested in it.
>
> If we were a new and used car dealer, or a company fleet manager, we might want to know other things about it, such as ownership, use and usage, options and accessories, etc. We might also want to know how many miles it has been driven, how much gas it uses, how many times it was serviced and how, how many times it was in an accident, how many different people have driven it, what it cost new, what it costs now, how much it costs to maintain, etc. These latter attributes are really process attributes. They are part of the description of the car, but these attributes tell us what was done to or with the car, not about the car as a thing.
>
> If we were an auto parts dealer, we might be interested in the parts of the car themselves, both new and used. In this case both the year, make, model, and color of the car become attributes of the part, along with its usage characteristics (if it is not a new part); its cost, size, shape, and weight; how many are used in a specific year, make, and model.
>
> A specific part could be elemental such as a bolt, tire rim, windshield, etc., or it could be a complex subassembly such as a transmission, radio, motor, etc. It could fit one year, make, and model of car, or any car. By combining several of these parts into a subassembly, we have in effect created a new "part."

All entities and most relationships have these types of process attributes associated with them. Process attributes are variable in that their values change frequently, and these changes usually involve the participation of some other entity. Thus, since they relate what one entity did to another, or where or how many of one entity are contained in another entity, they are normally descriptive of the relationship between the two, rather than descriptive of one entity or other, although obviously they could be.

The process of identification, definition, and contextual placement of the entities is vital to any understanding of the business and to any effort directed at either application development or file design. Processes like data normalization (a much-discussed concept) cannot be meaningful unless we know what those entities are, what the difference is between an entity and an attribute of an entity, and further what relationships exist between those entities.

This process of identification, definition, and contextual placement is greatly assisted by the creation of entity-relationship diagrams as one proceeds from analytical level to analytical level.

Entity-Relationship Analysis

Entity-relationship analysis is a multilevel process where each level produces a clearer and better defined view of the environment. The complete work products of this analysis result in a series of environmental definitions along with a diagrammatic representation of that level.

Enterprise level

The first, or enterprise, level consists of identifying the major entities of the firm. Although an entity is usually represented as a single instance, at this level each entity represents a whole class of people, places, or things. Here the definition is very general and represents all people, places, or things which relate to the firm in the same general manner, or which are viewed by the firm in the same manner.

Potential entities at this level might be employee, customer, organizational unit, order, or product. Each of these major entities should be related to at least one other major entity. The definition of the entity at this level does not distinguish between entity types, entity roles, or entity activities. There is no differentiation between the various subcategories of the entity nor are any other distinctions made.

The definition is made as general as possible without losing the concept itself. For instance, a customer may be defined as "any person or organization which buys, rents, leases, or otherwise acquires product or services from the company." A product is "any physical thing or service which the company provides to its customers in the course of conducting its business (not necessarily for a price)."

Every entity should be related to at least one other entity. There is no differentiation between the number or types of relationships between any two entities. Like the entities themselves, it is a binary condition: it exists or it does not.

The enterprise-level analysis has the broadest scope and the most general definitions of all the analytical levels. It is intended to serve as a broad brush view of the corporation and as a road map for further analyses. As the analysis proceeds to each successive level this model may be modified, with entities or relationships being added (the more usual case) or deleted.

An enterprise level analysis will usually identify from 10 to 30 entities. The number of entities which appear will depend upon the complexity of the corporation and the type of business.

Another determinant is the way in which the entities are defined. A definition of the general entity "employee" will be less complex than one which defines more specific entities such as sales, production, and back office. Again, a definition of the "organizational unit" as a general entity will be less complex than one which defines sectors, groups, divisions, or subsidiaries.

There are no rules governing how the entities are defined at this or any other level, except to say that the definitions should be consistent with respect to their level of abstraction (that is, do not define one entity as "all employees" and at the same level differentiate between all different types of products or customers). These definitions should also make sense within the context of the firm, and they should be as specific or as general as is necessary to make the diagram clear and readable.

Entity-relationship level

The second, or entity-relationship, level is an expansion of the enterprise level. At this level the entities which were previously identified only at the class level can now be brought into sharper focus. This level recognizes differing types of entities.

As with the previous level, the entities are in reality groups or sets of people, places, or things; however these groups may be much smaller and much narrower in definition than they were at the enterprise level. For instance, it may be relevant to a company to differentiate between types of customers, such as between institutional and retail in the brokerage industry; between subscription and mail-order in the publishing industry; between different types of products, such as spare parts and finished items, or between elemental parts and subassemblies.

In some cases the various entities may be differentiated by the role they play with respect to the firm. For instance the analysis might make distinctions among executive managers, middle managers, clerical workers, professionals, and sales personnel, between full-time and part-time workers, between salaried and hourly workers, or between union and nonunion workers.

Distinctions can be made with respect to functions, e.g., among production, sales, engineering, and back office personnel. In each of these cases, although the people in each category are all employees, they are treated differently by the firm or they play different roles with respect to the business of the firm.

As each distinct subclassification of entity is identified, it should be named. As with the enterprise level, each of the entities identified at this level should be related to at least one other entity, and may be related to many other entities. At this level each specific relationship which exists between each pair of entities is identified and named.

Except for recursive relationships (those where individual instances of the entity are related to other entities of the same type), each relationship should be between two entities of different types. In addition each relationship must have a name that is descriptive of the particular relationship between the two entities being joined.

Entity-relationship-attribute level

The third, or entity-relationship attribute, level is similar to the second-level diagram with the exception that attributes are identified for both the entities and the relationships. Each attribute is named, and that name is indicative of the type of information which that attribute represents. For a person entity these attribute names might be family information, residence information, physical description, hobbies, clothing sizes, etc.

These attribute names for an employee entity might be very similar to the section headings on an employment application, or the section headings on the permanent employee record form. For a customer, they might be very similar to the section headings on a form for opening a new account or a customer record.

Data element level

The final, or data element, level is one which is most familiar to data processing specialists. This level consists of identifying and defining the specific data elements which are needed to describe each attribute of each entity and each relationship. *Data elements are assigned only to attributes.* Since we know what the entity represents and we know what the attribute represents, the addition of elements is normally a relatively straightforward process.

Chapter

8

Modeling and Diagraming Techniques

CHAPTER SYNOPSIS

Much has been written about the modeling process, and in fact most prominent development methodologies incorporate modeling in some form or other. Implied in each of these methodologies is that their models can be applied to the entire analytical process. In our view, this is not the case. Each of these differing modeling techniques is aimed at a different portion of the analytical process.

The analyst should be familiar with the major modeling techniques and with their most appropriate application. Because it is one of the newest and least-documented modeling techniques, we will give particular emphasis to the entity-relationship model. This chapter will address some of the other modeling approaches and provide some simple, easy-to-use guidelines and procedures for building the various major models.

Modeling and Diagraming Techniques

A definition

A *model* is a representation, either graphic, narrative, or both, of a physical or conceptual environment. It must identify the major components of the environment, describe those components in terms of their major attributes, depict the relationships between the components, and describe the conditions under which the components exist and interact with each other. A model should depict in graphic and narrative form, the entities and their relationships within the application environment. A model can be composed of several independent or interdependent submodels.

Much has been written about the modeling process; in fact most prominent development methodologies incorporate modeling in some form or other. Each of these methodologies implies that their models can be applied to the entire analytical process. In our view, this is not the case. Each modeling technique is aimed at a different portion of the analytical process. Some models are most applicable to modeling data, some to modeling processes, some to modeling work flows, and some to modeling the decision-making process. Attempts have been made to modify some of these models for areas of analysis other than those for which they were initially designed. By and large these attempts have not been overly successful. The analyst should be familiar with the major modeling techniques and with their most appropriate application.

The Entity-Relationship Model

One of the newest modeling techniques, and one of the most powerful, is the entity-relationship model, or entity-relationship diagram, which is the modeling technique employed by the Entity-Relationship approach to analysis and design. The majority of the current literature references concentrate on the associated modeling technique, which is the heart of the methodology. Moreover, these references concentrate on using the approach for building data models. These data models, because of their orientation, are uniquely suited to the development of database logical models in the hierarchic, network, and relational environments. The entity-relationship model is equally suited to all three because of its real world approach to data.

The 1976 paper in which Dr. Peter Chen described the analytical basis for the entity-relationship approach also included a description of the modeling technique which is an integral part of the method. The multilevel analysis portion of the entity-relationship approach produces a series of environmental definitions, each one of which is accompanied by a diagrammatic representation of that level. These diagrams are simple, clear pictures of the environment in terms which any user can understand. In fact user input is an integral part of the diagram creation process.

Entity-relationship models are not data structure models. And, although at their most detailed level they contain and identify data elements, they are not data processing models. They are business models, and as such, they model business environments and depict business components.

Entity-relationship diagrams (also referred to as models) consist of representations of the various levels and parts of the organization, from the strategic to the operational level. Each model of a level represents the entities and relationships from the perspective of that level, and within a level the entity-relationship models represent the perspective of one or more particular users at that level.

Although there are numerous variations of the entity-relationship approach model notation, the three basic notational components of the entity-relationship model are symbols representing an entity, a relationship between two entities, and the attributes, or descriptors, of either entities or relationships.

These symbols are (see Figure 8.1)

1. *Rectangles.* Each unique entity type or subtype is represented by a rectangle which contains the name of that unique entity type or subtype.
2. *Diamonds.* Each relationship which exists between any two different entities or between two occurrences of the same entity is represented by a diamond which contains the name of that relationship.
3. *Circles.* Each unique attribute of either an entity or a relationship is represented by a circle which contains the name of that attribute.

Entity-relationship (ER) models have been applied to individual business units, and even to individual business functions. The full ER

Entity name

Entity symbol

Relationship name

Relationship symbol

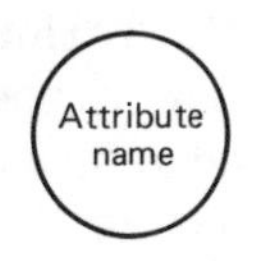

Attribute symbol

Figure 8.1 Symbols for entity, relationship, and attribute.

approach model addresses the whole organization and each of its parts in a top-down manner. Only by using this top-down approach with levels can the full business perspective be attained. The approach develops the models in pyramid fashion, beginning with the whole firm and proceeding downward. This corresponds most closely to the manner in which most firms view themselves.

We will present the models in the sequence in which they are most easily developed, a sequence which corresponds to the three levels of the organization: strategic, managerial, and operational.

The entity-relationship approach produces a different type of diagram or set of diagrams for each of the three basic organizational levels.

The enterprise level

The entity-relationship level

The entity-relationship-attribute level

Because it is a top-down approach, the contents of the diagrams at each successively lower level represent a decomposition or expansion of detail of the level immediately preceding (see Figure 8.2). The number of diagrams at each level is dependent upon the number of entities and relationships involved and on the complexity of those entities and relationships. There is no requirement that diagrams be maintained on a single chart, or that they have to be broken down into many smaller charts.

Aside from the enterprise-level model, which should be a single chart, and by definition a firmwide chart, the lower level charts may be developed against any perspective. These perspectives may be firmwide, or by function, business unit, or product line. Because they are designed to be an aid for analyzing and understanding the business environment, the diagrams at each level can be combined or split in any manner which aids comprehension, but above all they should be drawn in such a way that they are easy to follow and meaningful to both analyst and developer.

The Enterprise Level

Level one, the enterprise level, consists of an identification of the major entities of the firm and an indication as to whether a relationship exists between them. There is no differentiation among the various subtypes of any given entity nor any indication of the number or types of relationships between any two entities. The entities and relationships are represented in binary state, that is they either exist or they do not. At this level only the major entity classes are named. By def-

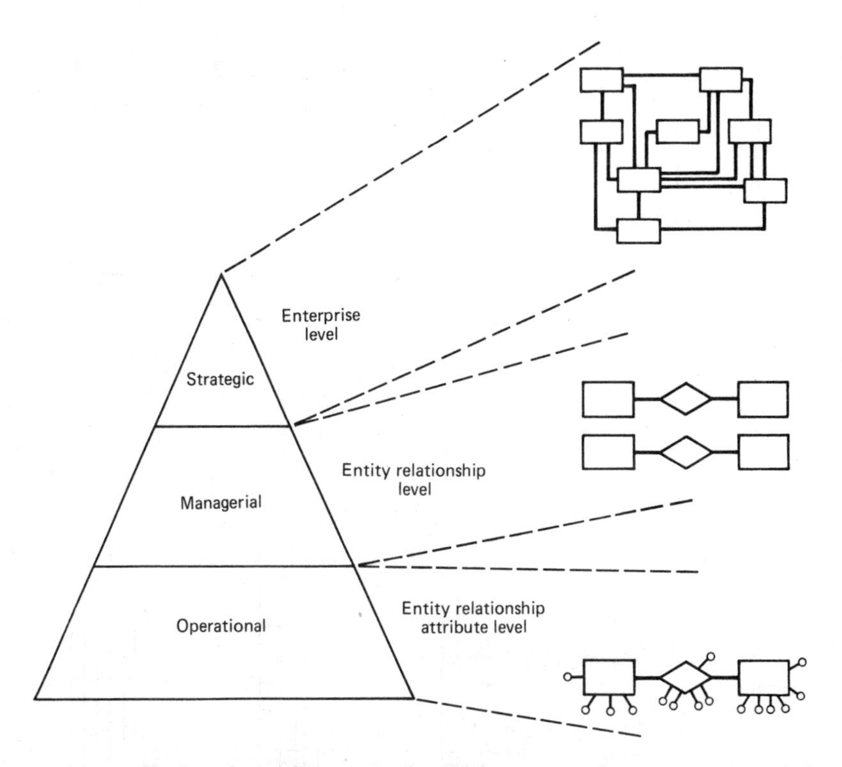

Figure 8.2 Entity-relationship approach models corresponding to organizational levels.

inition there can only be one enterprise-level model. See Figure 8.3 for an example of an enterprise-level model of a brokerage firm.

Every firm, large or small, deals with many different types of entities in the course of conducting its business. Although the names of the various entities will vary from firm to firm, at the most general level they can be grouped into four major categories: people, places, physical things (such as a document, a product, a machine, etc.), and logical or legal things (such as a corporation or a business unit). Using these four major categories, we can identify some of the most commonly occurring entities regardless of the type of business a company does.

People entities fall into three major classes: (1) the people who make up the firm's work force, (2) the people who are its customers or clients, and (3) the people who supply it with raw materials, products, parts, and financial or other services.

Place or location entities also fall into three major classes: (1) the places where its services are offered, or its products are made, stored, and/or sold, (2) the places where its work force is located, and (3) the places where its customers reside or are otherwise located.

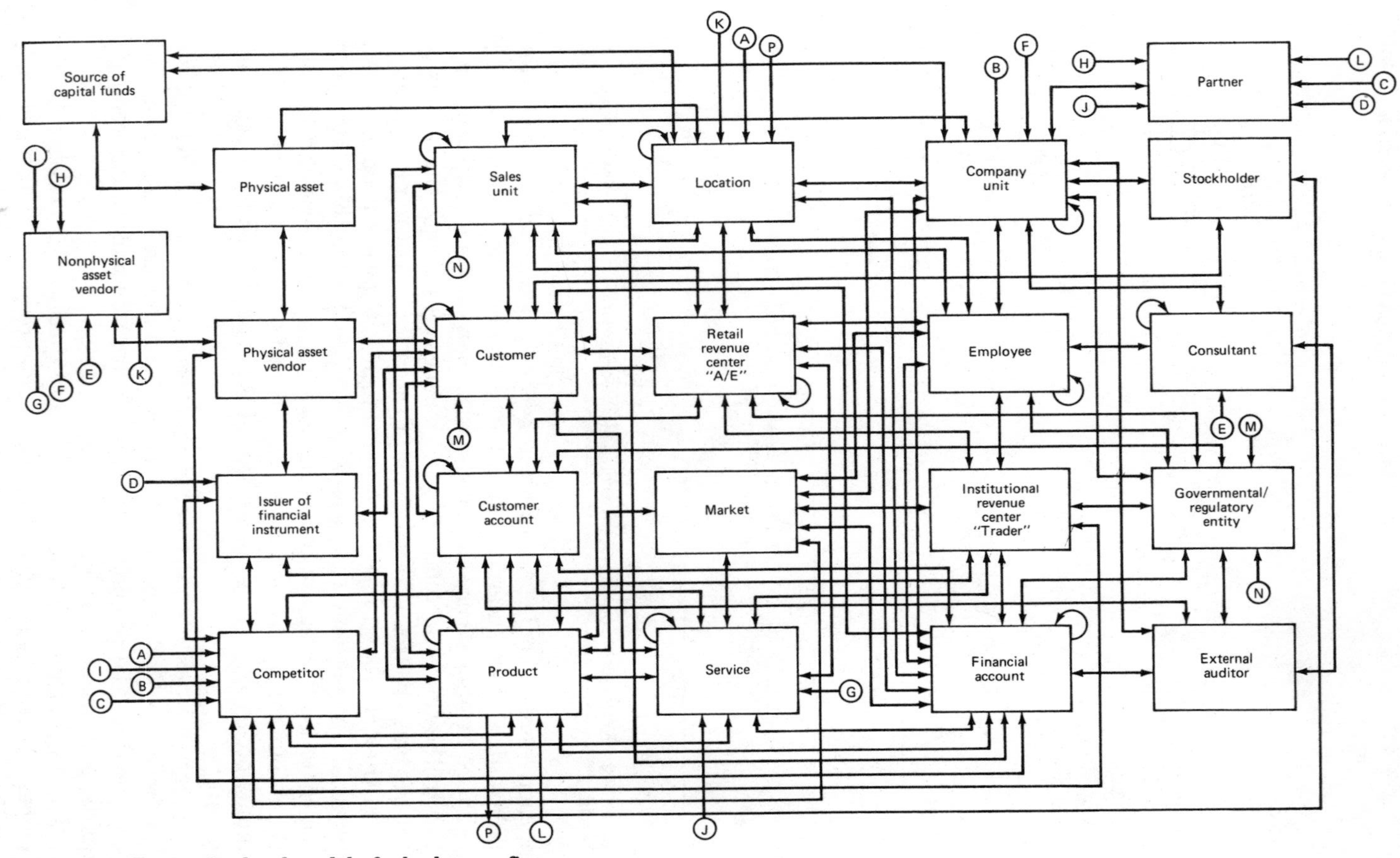

Figure 8.3 Enterprise-level model of a brokerage firm.

Physical thing entities include: the actual products of the firm; its physical assets (buildings, land, furniture, machinery or other equipment, inventory, supplies, etc.); its financial assets (money, securities, leases, contracts, bank accounts, loans, notes, credit lines, etc.); and the documents, memoranda, accounts, contracts, orders, invoices, statements, checks, vouchers, reports, and files which record its business transactions and activities.

Logical or legal thing entities include: the services offered by the firm; the firms that are its customers or clients; the firms or people who supply it with raw materials, products, parts, and financial or other services; the markets within which the firm operates; the governmental and regulatory units under whose jurisdiction the firm operates; and the organizational units into which the firm's work force is grouped for business, functional, and reporting purposes.

Within any given firm it can be expected that most, if not all, of the above entities will be represented. What they are called, how they are defined and described, how they are subtyped, and, more important, what the firm needs to know about them depends upon the specific business of the firm; its culture; and the business rules, policies, mission statements, charters, and procedures which govern what it does and how it operates.

It should be remembered, however, that the business entities being identified and defined at the enterprise level relate to the firm as a whole. As such these entities and relationships may be numerous and complex and may lack the precision of definition which lower level models require.

The enterprise business model depicts entity classes. At the class level the entities have the widest possible definition and scope while still maintaining the general physical and role characteristics of the individual entities of which they are composed. These entity classes are treated as if there were no variations in type and as if each of their component entities were defined and behaved in a similar manner.

At the enterprise level

- Entities are identified and named.
- Relationships are defined as either existing or not existing between any given pair of entities.
- All entities and relationships are viewed from a single perspective.
- Business rules stated are at a strategic or policy level and apply firmwide.
- Business activities are stated as functions.
- Business entities are portrayed at a class or universal level. There is no differentiation between the various subtypes of a given entity,

unless those differences have meaning at a firmwide and a functional level.

The creation of the enterprise-level diagram is a two-step process. Step 1 is to identify, select, and name the relevant entities. It is helpful, although not mandatory, to select a primary or core entity to begin the model. In most cases this core entity will be either the customer, product, or employee, or all three, since normally these are the most important entities to the firm. These entities are placed in the center of the page.

All entities directly related to that core entity are placed, one at a time, in a ring around the core entity. They will be called secondary entities only for purposes of describing the model creation process. As each entity is drawn, its name should be placed within the entity symbol. This is the entity's primary name (the name of all entities in the class) or its role name (the name of the entities in the subclass being depicted).

Step 2 is to connect each pair of related entities with a single line between them. This line contains no name or other information. Although a specific entity may be related to many other entities, each entity must be related to at least one other entity within the model.

The Entity-Relationship Level

Level two, the entity-relationship level, is an expansion of the enterprise level. At this level the entities which were previously identified only at the class level can now be brought into sharper focus. This level recognizes and names both the different subtypes of the major entities and the various distinct relationships which exist between them.

All models below the enterprise-level business model are more detailed and may use any of a number of distinct entity subtypes or subsets of the universal entity set in place of the universal entity set. Here, each subset is given a name corresponding to either the entity subtypes which populate it or to the role which the subset member entities play within the firm.

These names are usually something other than that of the entity name assigned to the universal set. These subsets are usually created to represent the various roles which the more global entity plays.

In some cases the subset name is different from the role name and may represent the title by which the members or principal members are known within the firm. In these cases, both the role and title names by which that entity is known should be stated.

At the managerial level

- Entities and entity subtypes are named.

- Each relationship between the pairs of entities or entity subtypes is identified and named.
- Entities and relationships may be firmwide or viewed from a variety of business perspectives, unit, divisional, or other organizational grouping. These perspectives may cross functional or business unit boundaries.
- Business rules are tactical and may apply firmwide or to a specific unit or set of units.
- Business activities may be function or process oriented.
- Business entities may be portrayed at a global or universal level or may be differentiated into more restrictive subtypes which relate to the particular business unit.

The creation of the entity-relationship level diagram is also a two-step process. Step 1 begins with the major entities represented at the class level on the enterprise-level diagram and differentiating them into their meaningful components, subtypes, subclasses, etc. It is here that the various types of customers are differentiated, as well as the various types of products, employees, accounts, etc.

Step 2 is to define each of the relationships between each primary entity and its related secondary entities. Each distinct identifiable relationship should be represented using a relational shape connected on one side to the primary entity and on the other to the secondary entity with which it is related. The relationship symbol should contain the name of the relationship type (Figure 8.4). Each secondary entity should be connected to every other secondary entity to which it is related in the same manner as the secondary entity was connected to the primary entity.

The above procedure should be repeated adding in third-, fourth-, fifth-level entities, etc., as appropriate. In each case, any significant relationships between each new entity and any previously drawn entity should be added as above.

The Entity-Relationship-Attribute Level

Level three, the entity-relationship-attribute level, is similar to the level two diagram with the exception that named attributes are added to both the entity and the relationship symbols.

For the entities, each attribute represents some grouping of data which is necessary from a business perspective to describe a physical or logical characteristic of the entity or to describe some activity of the entity. For the relationships, each attribute represents some grouping of data which is necessary from a business perspective to describe, qualify, or maintain the named relationship between two entities.

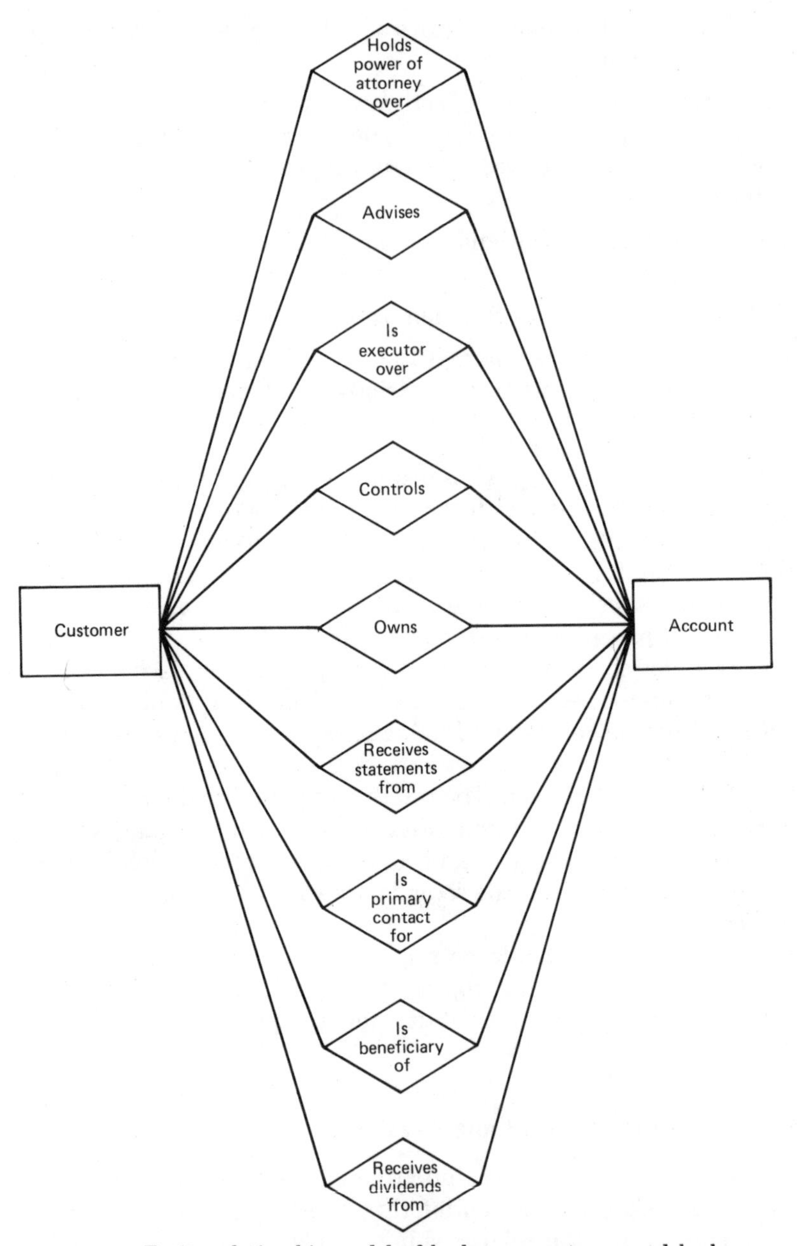

Figure 8.4 Entity-relationship model of brokerage customer and brokerage account.

- At the operational level
- All attributes of each entity and entity subtype are named.
- All attributes of each relationship between each pair of entities or entity subtypes are identified and named.
- Entities, relationships, and business activities may be firmwide or viewed from the individual perspectives of the operational, user, or application areas.
- Business rules are tactical and apply to a specific unit or set of units.
- Business activities may be oriented toward a functional process, an activity, or a task.
- Business entities may be portrayed at a global or universal level but are more probably portrayed as individual subtypes which relate to the specific operational unit.

The creation of the entity-relationship-attribute diagram uses the entity-relationship diagram as its starting point. Until this point the models have only identified the entities and relationships by name and context. For a given entity or relationship little is known about it other than its name and the obvious fact of its existence, and the fact that the firm is interested in it.

At this third level, we describe these entities in terms of their attributes or characteristics. In other words, beyond knowing that the entity exists, we must also know what the entity looks like, how it is identified, and what it does. These descriptors or characteristics are called attributes. An attribute is thus some distinct aspect of the entity or relationship and is necessary to describe the entity or to qualify the relationship. The full description of an entity or a relationship consists of the full set of attributes which describes it.

For an attribute to be significant it must relate directly to the entity or relationship, be completely dependent on the entity or relationship for its existence and meaning, and be definable in terms of one or more data elements. It is immaterial as to whether one or more data elements exist in an attribute, as long as the attribute applies to all instances of the entity or relationship being represented. Seen another way, an attribute is some distinct category of mutually related data, the sum of which describes something of interest about the entity or some qualifier about the relationship between two entities.

The creation of an entity-relationship-attribute diagram is a multiple-step process. Step 1 begins with the extraction of each entity from the ER diagram and its placement on a separate page. Each identifiable attribute of that entity that is represented by an attribute symbol should be drawn below the entity symbol and connected to the entity by a single line.

As each attribute is drawn, its name should be placed within the attribute symbol. As each attribute is identified and named, it should be annotated with a discrete number or n (denoting some unknown number greater than 1) to indicate how many occurrences of this attribute would necessary to describe the entity (Figures 8.5 and 8.6).

Entity attributes represent

- A physical characteristic of that entity—size, shape, weight, or color
- A historical attribute—date of birth or date of hire
- A locational attribute—place of residence, place of work, or place of birth
- A nonphysical characteristic—price
- An identifier—name or title
- An occupational characteristic—current position, skill possessed, training received, or educational courses, etc.
- The intermediate or final results of some processing activities related to the entity
- Data which relate to some current state or condition of the entity, or to some past or future state or condition of the entity
- Data which relate to some current action taken by or against the entity, or to some past or future action taken by or against the entity

At step 2, each distinct relationship between each pair of related entities should be extracted from the master diagram and placed on a separate page.

The attributes of each relationship are those categories of data which are necessary to qualify the relationship and describe when and under what conditions it occurs, and any other information which relates only to the connection between the entities and not to either entity independently.

As each attribute is identified and named, it should be drawn below the relationship symbol and connected to the relationship by a single line. As with the attributes of entities, the attribute should be annotated with a discrete number or with n, to identify the number of occurrences of this particular attribute which are necessary to fully describe or qualify the relationship (Figure 8.7).

The relationship attributes should include all attributes necessary to clearly and completely identify the many qualifications of that particular relationship between the two entities and the conditions under which the relationship exists. Relationship attributes (Figure 8.8) represent some descriptor or qualifier of the relationship such as

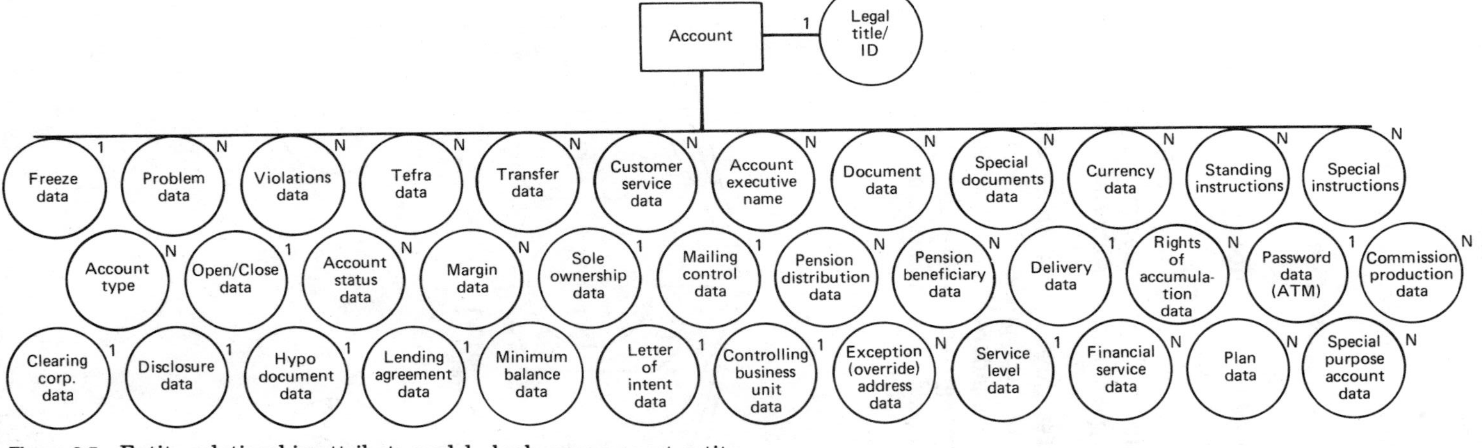

Figure 8.5 Entity-relationship attribute model—brokerage account entity.

- A historical attribute—date of marriage, date of sale, or date of storage.
- A locational attribute—place of storage, place of work, or place of birth.
- A nonphysical characteristic—price at time of sale, discount at time of sale, or grade in course.
- Some meaningful data which are not attributes of either entity participating in the relationship, but pertain only to the relationship between them. This type of data is sometimes called intersection data.

At each of the first three levels, these diagrams consist only of shapes with names. In data processing terms, and in a very general sense, they could be considered to be the identification of the record types (or record groupings) which will ultimately contain the data elements. It must be noted that the attributes represent mutually ex-

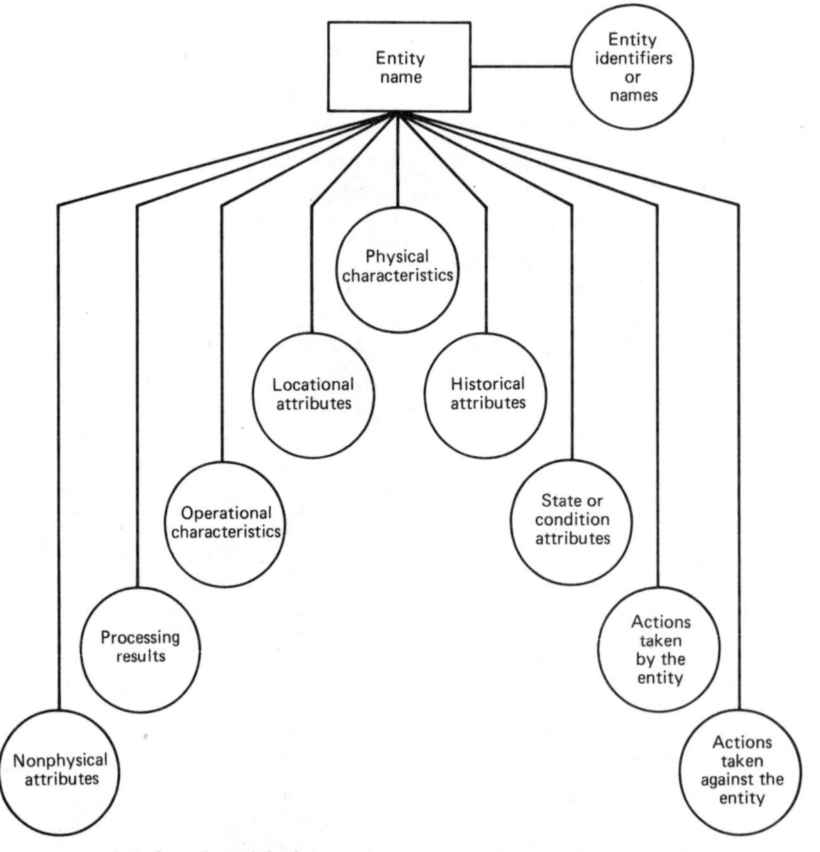

Figure 8.6 Attributed model of an entity.

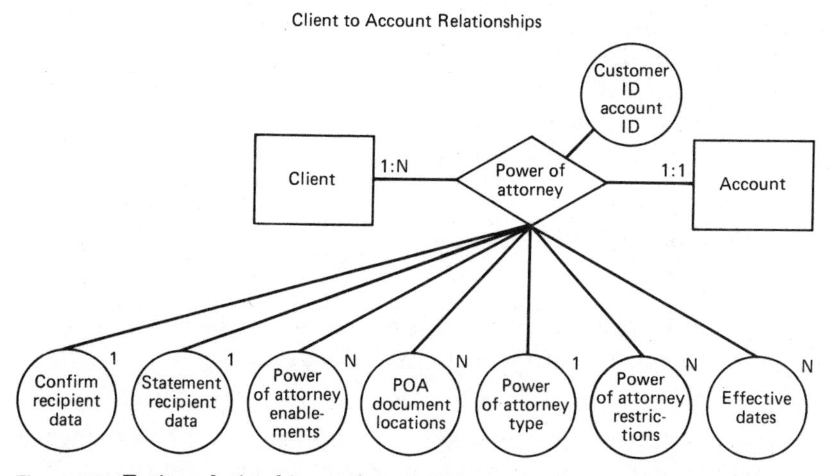

Figure 8.7 Entity-relationship-attribute model—power of attorney relationship.

clusive and mutually independent categories of data. They may or may not represent actual record types.

In the logical data structure models created at a later date from these ER models, attributes may be combined to form records or more general records, or they may be kept separately. The entity shapes are the names of the logical data aggregates (or structures) of the environment.

The Entity-Relationship-Attribute-Data Level

A fourth, or data element, level may be added when the models are developed in conjunction with the data processing systems development projects. This is the level which is most familiar to data processing specialists and consists of identifying and defining the specific

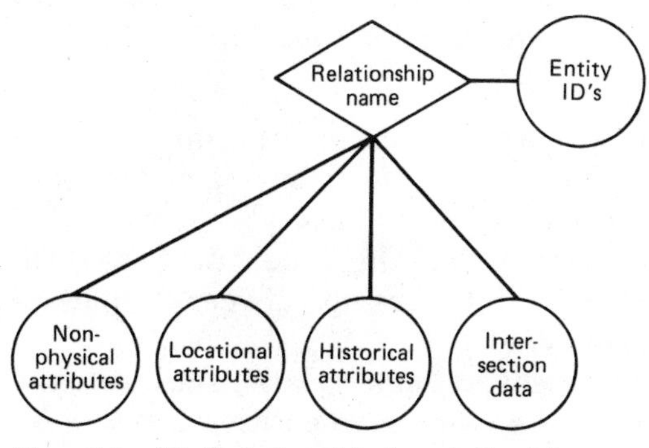

Figure 8.8 Attributed model of a relationship.

data elements needed to describe each attribute of each entity and each relationship. *Data elements are assigned only to attributes.* In a sense, data elements are the attributes of the attributes. Since we know what the entity and what the attribute represent, the addition of elements should be relatively straightforward.

Construction of an Entity-Relationship Diagram

Regardless of the level being addressed, the following rules apply to the construction of an entity-relationship diagram.

1. Entities
   a. Each rectangle must represent a single entity, a homogeneous group of entities, or one subset or subtype of the entity.
   b. When developing detailed models, each identified global entity should be decomposed into its component subsets.
   c. The mode of decomposition is dependent upon the characteristics of the component entities and upon the requirements of the firm for information about those entities.
   d. Regardless of the mode of decomposition, care should be taken to ensure that all entity subtypes can be related back to their base global entity. This may be accomplished by special notation or by including the name of the base entity within the entity subtype name.
   e. When the model includes documents, each unique type of document should be included, and the attributes for these documents should include all data field content which must be validated, processed, and retained by the firm.
   f. When document processing requires that data be validated against preexisting reference files, code lists, spreadsheets, or other financial tables, the referenced data items should be treated as if they were entities and included in the model along with their appropriate attributes and relationships.

2. Relationships
   a. Relationship diamonds are drawn between, and must be connected to (by a line from each side of the diamond), no less than one and no more than two entity rectangles.
   b. A diamond may be connected back to the same entity, in which case it represents a recursive relationship between unique occurrences of the same entity.
   c. Each diamond must represent a single relationship which is known to exist between the two connected entities *and is of interest to the firm.*

d. For each line which connects the diamond to a rectangle: At the point where that line joins that rectangle, a notation should be made as to whether the two entities being related have a one-to-one, one-to-many, many-to-one, or many-to-many relationship.

e. This notation should be made in the form $a{:}b$, where

a = the entity on the left side of the diamond
and b = the entity on the right side of the diamond

and a and b may have any numeric value equal to or greater than 1, or N (denoting an indefinite number more than 1).

f. A relationship is symmetrical if entity a (the left-hand entity) has the same relationship to entity b (the right-hand entity) as entity b has to entity a, or, in other words, each a is connected to many b's and each b is connected to one and only one a. If a relationship is symmetrical, then the notation closest to each entity should be the same.

g. A relationship is asymmetrical if entity a does not have the same relationship to entity b as entity b has to entity a, or, in other words, each a is connected to one and only one b, but each b may be connected to many a's. If the relationship is asymmetrical, then the notation closest to each entity should reflect the view from that entity to the opposite entity.

3. Attributes

a. Circles representing attributes are connected to either rectangles or diamonds.

b. Each circle may be connected to one and only one rectangle or one and only one diamond and must represent a specific attribute of the entity or relationship to which it is connected.

c. The circles on the diagram contain a name which identifies the specific attribute or set of attributes being depicted.

d. The line connecting the circle to the entity or relationship should be annotated to reflect whether the named attribute may occur only once per entity (or relationship) or many times.

e. Each entity rectangle and relationship diamond must have at least one associated attribute.

f. Attributes which apply to more than one entity, or to more than one relationship, must be diagramed as if they were unique to each entity or relationship to which they apply. This condition will occur when a global entity has been separated into entity subsets and the members of one or more subsets share many of the same attributes. Under these conditions each occurrence of the attribute symbol should have the same name, and some notation which indicates that it is identical in format to attributes which appear elsewhere.

If all attributes and all relationships connected to the entity rectangle do not have the potential to apply equally to each and every entity occurrence defined to it, then the definition of the entity being used must be changed and a new entity, new entity set, new entity subset, or new entity subsets, must be created until this condition is satisfied.

The above discussion assumed that one and only one model will be created at each level for the firm. However since most projects are undertaken for specific user areas, it may be desirable to create different models for each user area.

Just as an entity can be viewed from many different perspectives, and may seem to be different from each perspective, so too entity-relationship and entity-relationship-attribute models can be different from the various perspectives of the firm. Each area of the firm defines the entities of the firm in different ways and relates to them in different ways.

All entity-relationship models need not contain every entity of the firm. They need only contain the entities of interest to the particular area being modeled. That is,

1. A model can be built to reflect only the document entities, the entity sources for those documents, and the relationships between both types of entities.
2. A model might contain functional entities and their relationships. Here the functional areas of the firm (managerial concepts) are treated as entities themselves and the model reflects their relationships to each other.
3. Another variation might contain only the processing entities (groups of people, machines, or workstations) and the document and/or resource entities used by them. This type of model might reflect all the processing stations through which a particular document must travel, or the workstations through which a manufactured part must pass. A process model does not reflect what processing is done, or even how it is done, but rather the stations where processing of a particular type is done. That processing could be complex or simple.

These types of models are business models, rather than data processing models. That is they reflect business environments not methods of processing. The types of entities and relationships selected to be included in each model, the definitions of those entities, and the attributes used to describe those entities and relationships all combine to define the environment being modeled and the nature of the model itself.

Documentation of the Model

As the analyst completes the individual diagrams, he or she should be preparing dictionary entries for each attribute of each entity and each entity-to-entity relationship identified and defined. A description of each relationship should be prepared and entered into the dictionary as well. The first time an entity is extracted from the master diagram, a description of that entity should be prepared for the dictionary. Although any given entity may appear on multiple diagrams, only a single entity description need be entered into the dictionary. Thereafter the entity description will need to be updated with any additional relationships which it participates in, or for any additional attributes associated with it.

Each attribute should be identified, at minimum, with its name and the list of data elements of which it is composed. Each entity should be described, at minimum, with its name, a list of all subset entities contained within it, all aliases by which the entity is known, the names of all the relationships which it participates in, and the list of all attributes which are used to describe it. Each relationship should be described, at minimum, by its name, the entities involved in the relationship, and the names of the attributes associated with it.

Data Flow Diagrams

The data flow diagraming technique is usually associated with Yourdon and DeMarco, its developers and primary proponents. Data flow diagrams depict data flows and data transformation processes. A data transformation process is one which transforms input data into some form of output data. The orientation of data flow diagrams is toward business processes, the data which feed them, and the data which they generate. In addition, data flow diagrams are also oriented toward the system level and result in system-processing specifications.

A data flow diagram utilizes three symbols: a circle, which represents a process; a curved line with an arrowhead, which represents the flow of data to or from the process along with the name of the data input or data output; and two short parallel lines representing a data store (Figure 8.9). In some instances a data flow diagram may also contain rectangles which represent terminators. A terminator is a data source or data end user.

Data input to a process may originate with a terminator, another process, or a data store. A data store is synonymous with a file and

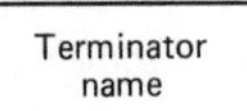

Figure 8.9 Data flow diagram symbols.

holds data between processes. In a data flow diagram, all processes and data stores must have at least one input and one output.

As with entity-relationship diagrams, data flow diagrams are constructed in stages, or levels. The first, or level zero diagram (Figure 8.10), is called the context diagram and contains a single bubble which represents the entire system. To that bubble are connected data flow lines representing all major inputs and outputs.

Once the context diagram is completed, a level one diagram should be drawn. A level one data flow diagram is called an essential model and depicts the major processes within the system. To construct it (see Figure 8.11), one first starts with a bubble which represents a major process. To that bubble are connected any input and output data flow lines associated with it. Each line is identified with the name of the particular data type. The process is identified with the name of the process which it depicts. Each input data flow line must originate with a process, a terminator, or a data store and terminate at the process bubble. All output data flow lines must originate with the process and terminate at a terminator, a data store, or another process.

As a second step the analyst should take each identified input and output from that process and add either the data stores, terminators, or other process bubbles from which the data originate or to which they are sent.

Figure 8.10 Level zero data flow diagram (context diagram).

The above process is repeated until all identifiable and significant data flows, data stores, terminators, and processes have been added to the diagram. The final step is to number each process bubble.

Once the above process is completed, the analyst may proceed to the next level diagram. Each process in the higher level diagram is decomposed into its more detailed component processes. For each of these more detailed component processes, each of its data flows, data stores, and terminators should be represented. These data flows, data stores, or terminators may be carried down from the preceding level, or they may completely originate and terminate within the major process itself. This diagram is completed when each of the major processes from the preceding level has been decomposed and diagramed.

The process of describing the data flow of each level is continued until the analyst arrives at a set of diagrams where each process represents a single task or data transformation with its associated input and output data flows and their associated data stores and terminators.

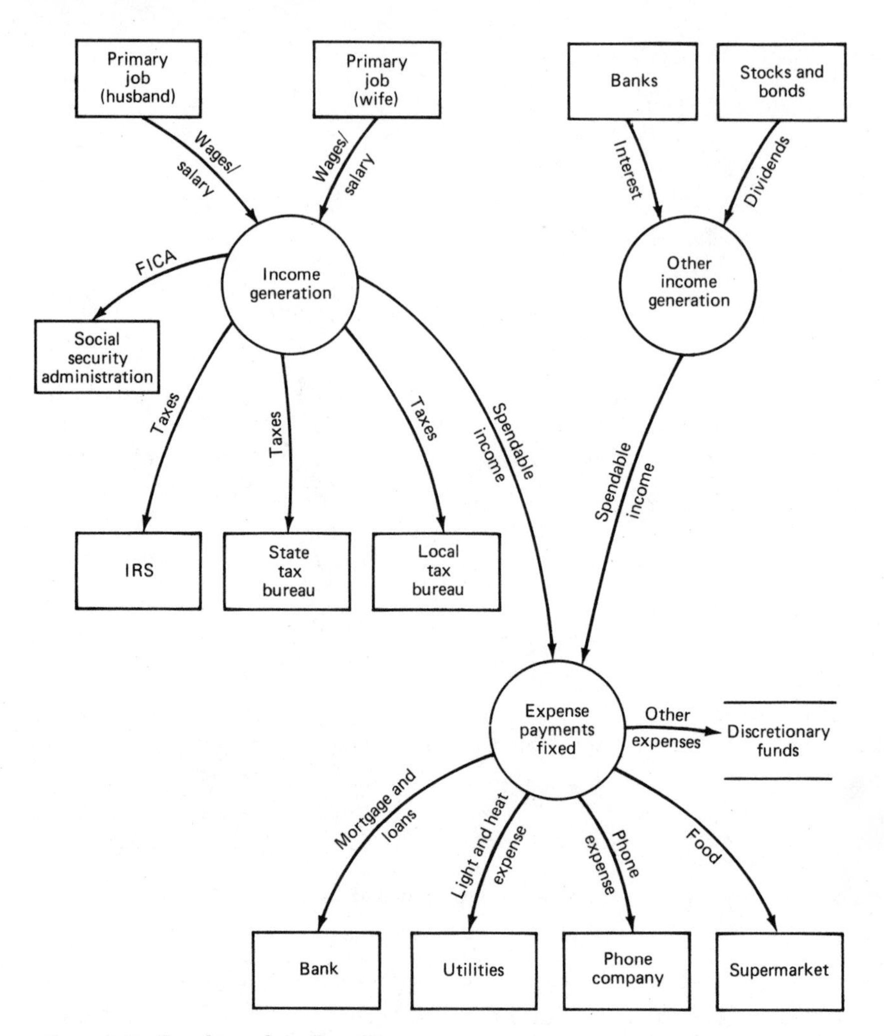

Figure 8.11 Level one data flow diagram.

The processes at each level should be associated back to the parent process at the preceding level by means of some scheme of numbers, or numbers and letters. All processes, data flows, data stores, and terminators should be described and entered into the analyst's data dictionary. Each data flow should be described, at minimum, with a name and a list of the data elements which compose it. Each process should be described, at minimum, with its name, and the tests made against the data, or the formulas which represent the transformations from input to output.

Flowchart Diagrams

Flowchart diagrams are useful tools when one wishes to represent either the processing and decision logic flows within a particular process, with its associated inputs and outputs, or the flow of an entire system, with all the various types of processing, preparation, inputs, outputs, data storage media, and other hardware which are associated with it. Although used primarily for depicting the implementation flows of data processing systems, it can be useful in an environment where one has a mix of both automated and manual processes, data stores, forms, and hardware.

Flowchart diagrams use a variety of symbols which singly or in combination represent the various data input and output sources, media types, and various processing and decision points and processing steps within a logical flow (Figures 8.12 and 8.13). Most data processing hardware vendors, and many software vendors provide plastic templates with the various basic symbols depicted in cutout form. These symbols may be used singly or combined with each other to form a large variety of symbols. Since there are a wide variety of different symbol sets in use, the analyst should clearly label each symbol used in a legend on the chart.

Flowchart diagrams that depict system flows are usually drawn in a linear fashion beginning at the top of the diagram, with either a manual operation, a form, a manual input or terminal input device, or a terminator which represents an end user. In some cases, flowcharts may begin with a tape or disk symbol. These symbols represent input sources and are connected to other symbols which represent either manual operations or automated processing points.

Each of these processing symbols is connected to some output media symbol, either a tape, disk, form, or display station. Each of these outputs is then connected either to another processing symbol as an input (with other first use inputs) or to inputs carried over from prior processing, or to a terminator, representing short- or long-term storage or an end user. This type of flowchart is complete when all processing boxes have been drawn and all outputs have been sent to storage or to a terminator (Figure 8.14).

Flowchart diagrams which depict processing and decision logic usually contain a more restricted set of symbols, usually process, input, output, decision, and terminator symbols. Figure 8.15 shows a simple logic flow diagram. In some cases they may also contain data arranging symbols (sorting, collating, etc). These flowcharts are also drawn in a linear fashion, starting with some input at the top. Since they depict only the tasks and individual steps within a process, they do not represent external data storage or entities. Input and output symbols are drawn only to depict at what internal step the data either enter or are generated.

Normally, these intraprocess flowcharts depict each separate step and decision point in the data transformation or usage process. Decision points are usually depicted as a diamond shape with each valid condition or test result indicated as a branch from the symbol. Each branch leads to a separate processing leg or sequence of steps. These processing legs or the sequence of steps may remain separate for the remainder of the process, may join the main stream later in the processing flow, or may terminate after some error-handling procedure. In some cases, they may loop back to some earlier point in the processing

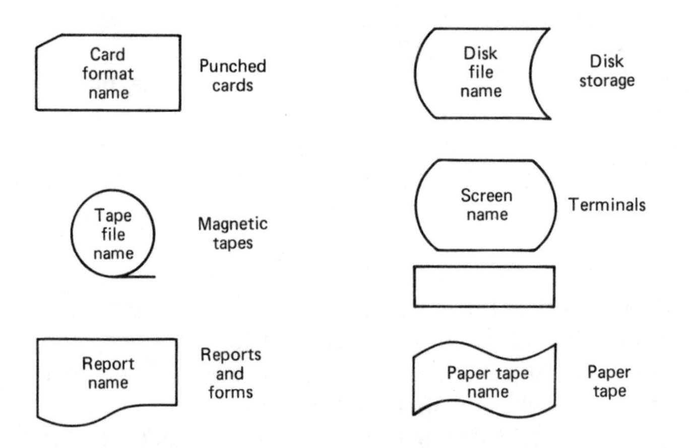

Figure 8.12 Data storage and input or output symbols.

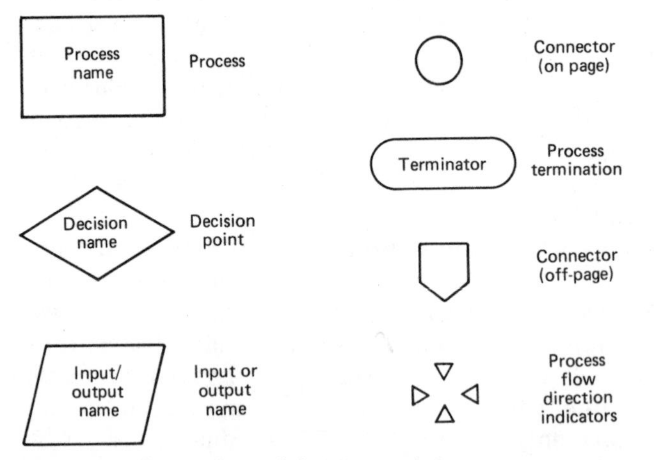

Figure 8.13 Processing and decision symbols.

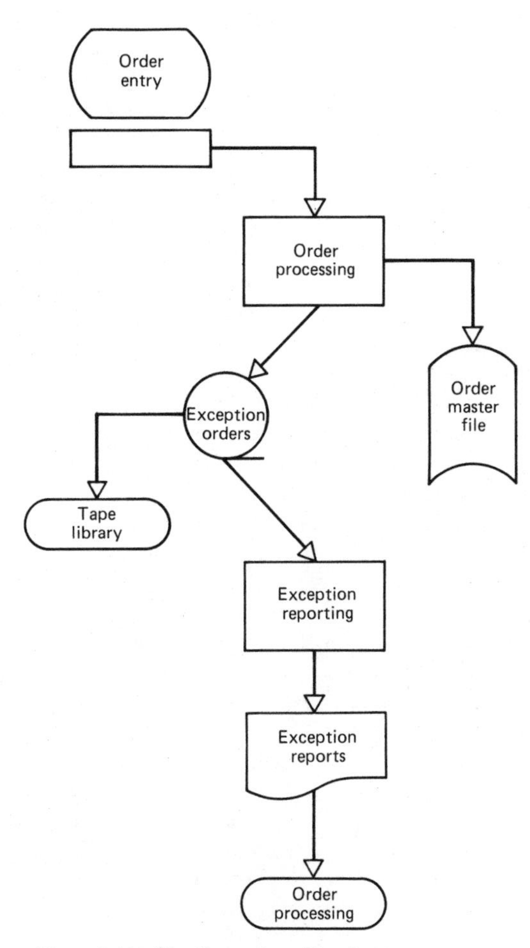

Figure 8.14 Simple system flowchart.

flow after corrective action has been taken or to remake the decision after additional processing has occurred.

Hierarchic Process Diagrams (Figure 8.16)

A process can be considered to be a series of activities, each activity consisting of a series of tasks. Although this view is easy to visualize, it is only partially accurate. In many environments, the tasks which constitute a process are not all performed every time. The particular tasks performed and the sequence in which they are performed is dependent upon the stimulus which activates the process. In some cases a task may be performed once, in others it may be performed multiple times, and in still other cases, it may not be performed at all.

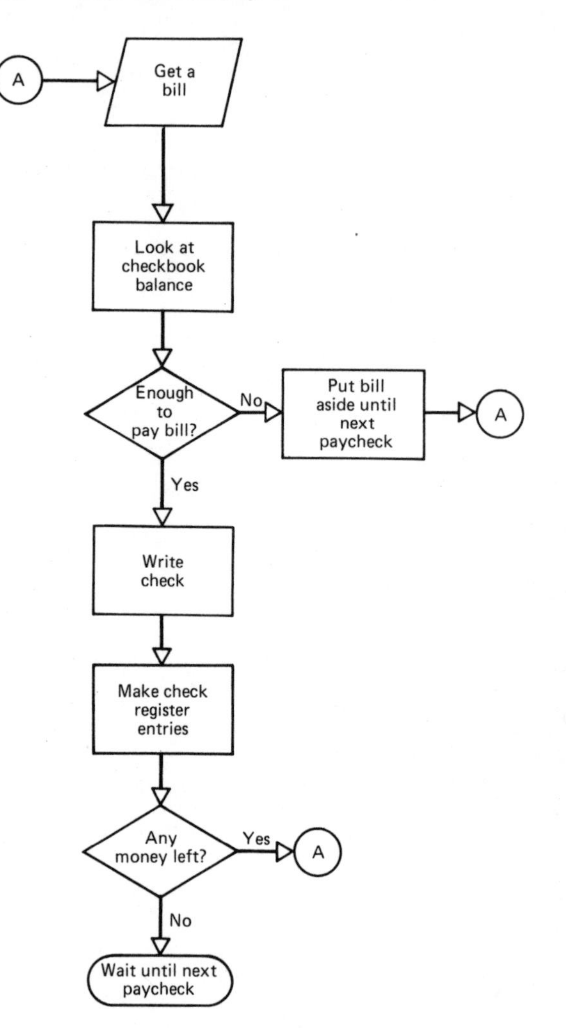

Figure 8.15 Simple logic flow diagram (pay bills).

Each task can be thought of as being dependent upon some preceding sequence of tasks. Viewed in this manner, the task sequence can be seen as a hierarchy with multiple levels and multiple legs. Rectangles are used to represent the tasks. In all cases there is a primary or root task, i.e., one which begins the process, or which must precede all others. Once this task, which is usually the one that receives the input document, is performed, it is usually followed by one or more second-level tasks which may be sequential or random, and mandatory, optional, or mutually exclusive. Each of these tasks in turn may trigger, or require the performance of, still other tasks which may also be sequential or random, and

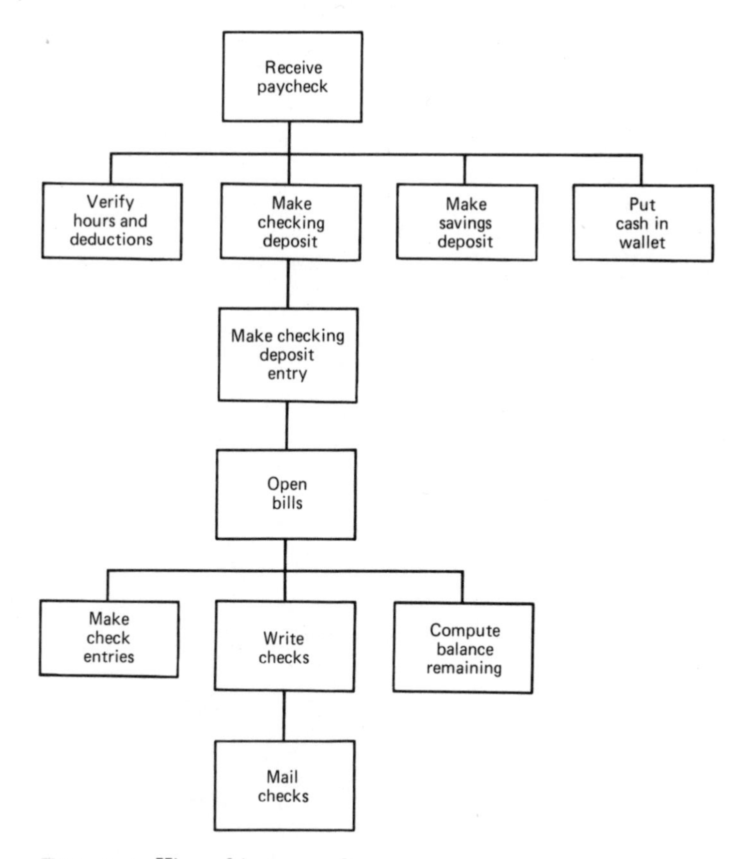

Figure 8.16 Hierarchic process diagram.

mandatory, optional, or mutually exclusive. This diagrammatic method is similar to a flowchart in that the flow of processes or tasks can be easily represented. It differs from data flow diagrams, flowchart diagrams, and entity-relationship diagrams in that data are usually not represented. Another difference is that logical decisions are not depicted.

Part

2

Analysis

Remember: The more things change the more they remain the same. Yet, the only thing constant in life is change.

Chapter

9

Business Knowledge

CHAPTER SYNOPSIS

One of the major goals of the analyst when starting any new analysis and design activities, regardless of whether the user area is familiar or not, is the acquisition of information about the current state of the firm environment, the user, what the firm does, and the place of the user and the user's organization within the body politic of the firm.

This chapter discusses the concept of business knowledge and its importance and relevance to the analytical process.

What Is Business Knowledge?

One of the major goals facing the analyst when starting any new analysis and design activities, regardless of whether the user area is familiar or not, is the acquisition of information about the current state of the firm environment, what the firm does, the user, and the place of the user and the user's organization within the body politic of the firm.

The term "firm" will be used to identify the organization as a whole. In some cases the term will represent a discrete business unit or division which can be examined as if it were a freestanding firm. This information, necessary for any analyst, is usually termed "background information." It can be obtained from experience in the firm, from the user, from user documentation, from a "briefing" by the analyst's manager or other knowledgeable source, or from a variety of similar sources.

Whatever it is called and however it is obtained, in reality this knowledge of the firm and user's role within that firm is called "business knowledge." Business knowledge is defined as a thorough understanding of the general business functions and the specific areas under analysis. The accuracy and completeness of this knowledge are crucial to the development of the foundation upon which the analyst can build an understanding of the user and the user's problems and requirements, and design new and more efficient methods of accomplishing the primary tasks of that user.

The task of gathering business knowledge includes acquiring a detailed understanding of the firm's functions, the processes and tasks which are employed to accomplish those functions, and the relationship of those functions, processes, and tasks to each of the functions, processes, and tasks performed throughout the firm.

The development of the analyst's business knowledge, regardless of the level at which the analysis is being performed, begins with the identification and description of each of the functions of the firm and of the user's function, or role, within the firm. This analysis, called functional analysis, usually includes a firmwide organizational chart and narrative descriptions of the functions and subfunctions which make up the business.

Generally the information gathered is collected into a set of documents which become the first stage of the analysis process. The collection of documents resulting from this process may be called a general business description document, a client functional description document, or another similar title.

The development of business knowledge is an integral part of the analytical process. In fact, the acquisition, documentation, validation, and evaluation of business knowledge is the core of analysis. By its very definition, business knowledge is knowledge about the business, what it is, what it does, why and how it does what it does, and how those activities can be performed more efficiently.

Business knowledge can be developed at any level; however, the higher the level at which the analyst begins, the more comprehensive and meaningful that knowledge becomes.

To use an analogy, it is possible to understand how an automobile works by studying its component parts: the engine, the transmission, the braking system, the steering mechanism, or the exhaust system. However to gain a better understanding of how an automobile works, it is more meaningful to understand how the various parts interact with each other; e.g., how the exhaust system ventilates the engine and how power is transmitted to the drive wheels by the transmission.

At a higher level, it is more meaningful to understand why each component is necessary and what the engineering principles are behind the internal combustion engine, the principles of physics behind the gearing in the transmission, and the chemical principles behind both the internal combustion engine and the exhaust system.

At a lower level, one can examine how the parts are put together and how the various subassemblies of the engine, the pistons, the carburetor, the manifold, the cooling system, the lubrication system, and the fuel feed mechanisms work.

Each approach is valid, and depending upon what you are trying to do, it may be possible to ignore all the other considerations and concentrate on the part which is faulty. For instance, one does not need to be an automotive engineer to change a tire or fill the fuel tank. However, if you were going to write a column on car repair or a book on automotive design, you would obviously need a greater depth of knowledge.

To use an earlier analogy, that of the medical profession, the surgeon specialist needs a deep background in anatomy, physiology, immunology, pharmacology, and related topics. The family doctor, while needing some background in these subjects, has less need in these areas but needs depth in diagnostic procedures and minor surgery. The nurse has less need than the family doctor for these subjects, but more need than the paramedic, and the least knowledge is needed by the first aid technician.

Analysis in a business environment deals with business components. These business components are the units into which the organization is divided, the functions of those units, the processes and activities associated with those functions, the tasks associated with those activities, and the data or information associated with those tasks.

If we can view the organization as an organism, then those organizational units are its internal organs, and their associated functions, processes, and tasks are their reason for being. Each unit and its activities serve a purpose. The data entering the firm can then be likened to the body's sensory inputs, and the information and decisions made by the firm are the thought processes based upon those inputs. Through its own actions the body uses the materials around it, gathered by it, or sent to it, as well as its thought processes, to produce other materials and products. The firm can be viewed in this manner as well. See Figure 9.1.

Just as the internal organs, sensory stimuli, and thoughts trigger actions by the body, so too data and the decision-making processes by the organizational units trigger business activities. Each body organ serves a purpose, and no organ operates independently of any other organ. So too, each business unit serves a purpose and no business unit operates independently. They are as interrelated and interdependent as the body's organs.

But again just as the body has some organs which have outlived their usefulness, which sometimes become diseased, or which cease to function properly, so too the organization has units which fall into the

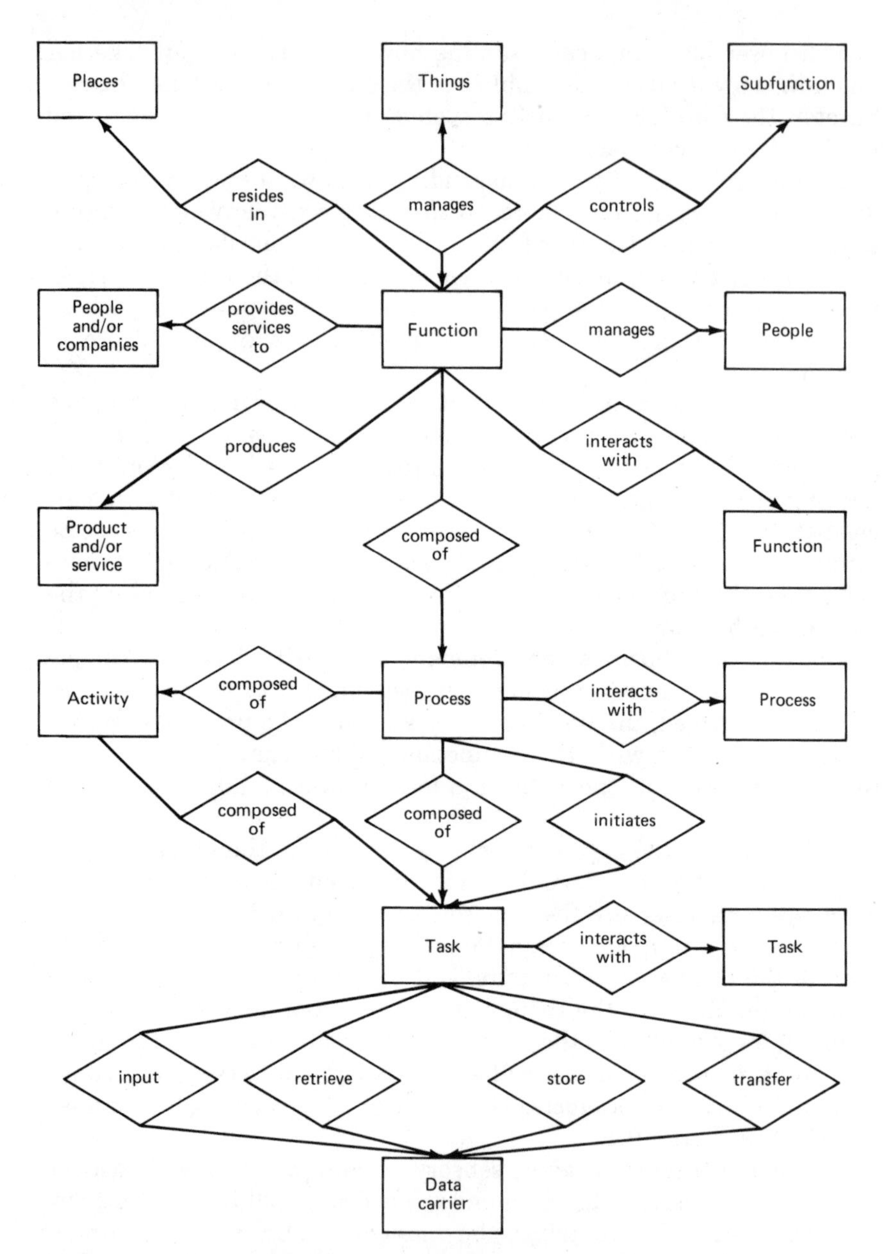

Figure 9.1 Entity-relationship model of function, process, activity, and task.

same categories. To extend the analogy a bit further, the body develops patterns of action and habits which, as time passes, may need to change, or which should not have been started in the first place. Sometimes external situations change, and the body must adapt its actions to those changes. Overeating, smoking, excessive reliance on drugs and alcohol, excessive exercise, or excessive dieting, to name a few, are all body actions which need to be changed.

Sometimes the person's relationships with others or their location need to change as well. Other changes are dictated by the aging process itself, such as the need to change diet because of disease, to wear glasses or a hearing aid, or to use artificial aids to move around.

The firm as an entity undergoes similar types of changes and a similar aging process. The analyst's role in the firm is similar to the physician's role with respect to the body.

But just as the body needs not only physical examination but also mental checkups and behavioral modification from time to time, all of which usually result in some change, so too the firm needs similar examination and modification if it is to remain healthy.

Just as the physician and behavioral specialists need a thorough understanding of the interaction between the physical body and its actions and motivations, so too the analyst needs a thorough understanding of the firm's "organs," actions, and motivations. Modification of the body and its actions requires an understanding of what the "correct" or "proper" actions or behavior should be and of what the body is capable of doing. This same logic and understanding also applies to the firm.

Bodily functions are fundamentally the same from person to person; yet each particular body is different, acts differently, and is differently motivated. So too firms are fundamentally the same in many respects, but each firm is different, acts differently and is differently motivated. The analyst must know and understand the fundamentals of the firm, and must uncover the differences in structure, action, and motivation. The process of obtaining this understanding, that is, the resultant uncovering of differences and problems, examination, and evaluation is the process of gaining business knowledge and analysis.

There are many modes of analysis and evaluation. Some are top-down, others are bottom-up. Some examine actions, stimuli, and environment and deduce motivation from them. Others aim at uncovering motivation and attempt to develop appropriate modes of action. Some work from stimuli to action, others from desired action to necessary stimuli.

As with our discussion of methodology, we will use the top-down approach and a motivation to mode of action approach. Again this is done with the recognition that there are many ways to accomplish the same result, but for purposes of discussion one needs to follow a consistent path to develop the ideas.

We will begin with a discussion of functional analysis and follow it by a discussion of process, task, and data analysis. We will follow a top-down general-to-specific approach, because we believe it is easier to take a complex object or idea and break it down into its component parts, than it is to build up complex ideas or objects from their component parts (see Figure 9.2).

Unless the reader is a watchmaker, in which case either mode is appropriate, it should be obvious that it is easier to take a watch apart and then draw a schematic of the watch, than it is to take a box of watch parts and try to build a watch from them while drawing the schematic at the same time.

If the watch does not work to begin with, unless one uncovers the symptoms and circumstances as to when it stopped working, the only way to fix it is to take it apart, look at all the pieces, fix or replace the defective items, and then put it back together again. If the process is done correctly and the break down is fully documented, you will fix the watch and there will be no left over pieces.

The larger, more complex, and unfamiliar the object, the more time must be spent in examination, documentation, and gaining understanding. We predicate the following discussion on the assumption that the analyst will probably not be familiar with the user business unit or its functions, activities, and tasks, and more probably will have only a vague understanding of the place of the user within the organization and an equally vague understanding of the structure, composition, and functioning of the business itself.

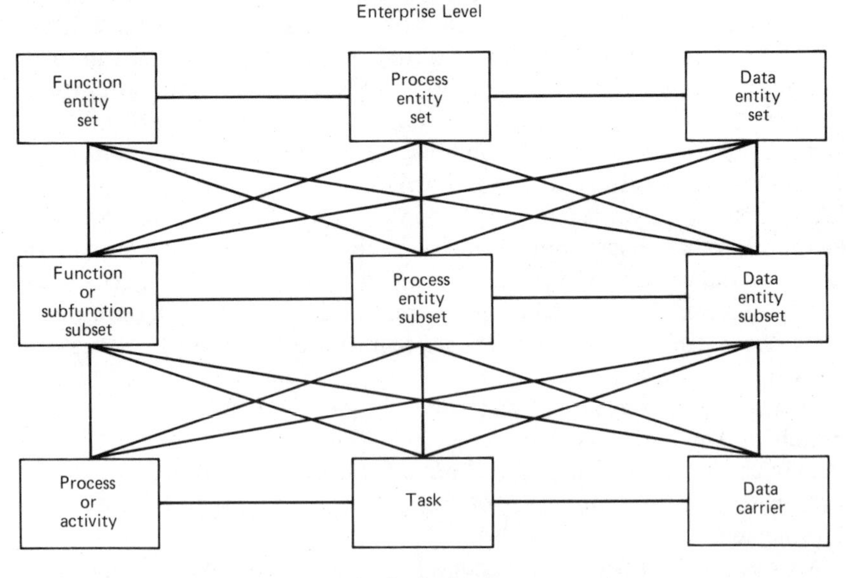

Figure 9.2 Function, process, and data analysis within the development cycle.

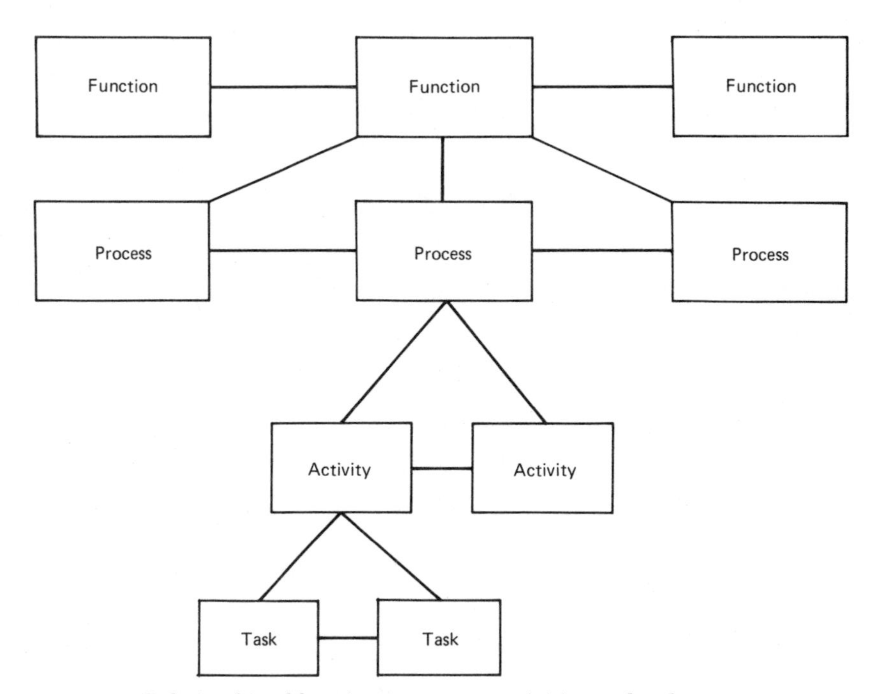

Figure 9.3 Relationship of functions, processes, activities, and tasks.

Analysis is a step-by-step process. It is relatively easy to discuss the process of gathering information, somewhat more difficult to explain the process of evaluation, and most difficult to discuss the process of arriving at appropriate recommendations. We will attempt to provide guidelines in the form of lists of questions, and discussions of the types of evaluation procedures which are available. We will present the advantages and disadvantages of various approaches where appropriate. The analyst should, however, realize that with most things and most situations there are no right and wrong answers, only better choices.

In the final analysis, the business is what the user says it is and the best approach is whatever the user will accept. The user is the final determinant of what is acceptable, and in most cases the user is the one who pays for the analyst's efforts to begin with.

The corporate body acts or reacts in many ways. The corporate structure is hierarchical with functions being at the highest level and tasks being at the lowest. In between these are systems, processes, and activities, which are all greater or lesser aggregates of specific task sequences. The terms themselves are usually used interchangeably. (See Figure 9.3.)

Chapter 10

Functional Analysis

CHAPTER SYNOPSIS

A function is a series of related activities, involving one or more entities, performed for the direct, or indirect, purpose of fulfilling one or more missions or objectives of the firm, generating revenue for the firm, servicing the customers of the firm, producing the products and services of the firm, or managing, administering, monitoring, recording, or reporting on the activities, states, or conditions of the entities of the firm.

This chapter discusses the concept of business functions and provides some insight into their identification and documentation. It includes extensive lists of questions for the analyst to ask when interviewing at the functional level.

What Is a Function?

A definition

A *function* is a series of related activities, involving one or more entities, performed for the direct, or indirect, purpose of fulfilling one or more missions or objectives of the firm, generating revenue for the firm, servicing the customers of the firm, producing the products and services of the firm, or managing, administering, monitoring, recording, or reporting on the activities, states, or conditions of the entities of the firm.

A function should be

- Identifiable and definable, but may or may not be measurable
- Definable in terms of activity, responsibility, and accountability

A function may

- Be a major area of control or a major activity of the firm
- Be composed of one or more subfunctions
- Be performed in one area or in multiple areas
- Be performed by an individual, a group of individuals, groups of individuals, areas of the firm, or the firm itself
- Involve one or more distinct, dependent or independent activities
- Be identified and defined without being performed

If a function is performed in multiple areas, those areas may or may not be related in either an organizational or reporting sense. A function is normally chartered to perform one or more related activities. These activities are usually related because they work on common data entities or because the activities are sequential or parallel and perform related work to a common end.

Generally, firms are organized along function lines, in that each vertical organizational grouping performs the same or related set of activities and is responsible to a single control point. Functions may be grouped into superfunctions which come together at very high organizational levels.

Two Categories of Business Functions

The functions of any firm, whether it is a manufacturing, finance, or service firm, can be segmented into two broad categories (Figure 10.1). Each of the various functions of the firm can be assigned to one or the other of these categories, and in some cases, a function may be assigned in both categories.

Business category

Functions in this category consist of those activities which are directly involved in producing the products, providing the services, generating the revenues and the profits of the firm, or managing those areas. This category has been termed the *operational* area of the firm. The functions in this category have also been termed *line* functions.

Administrative, support, or overhead category

This category contains those functions which service the firm as a legal entity and provide for its day-to-day well being. The administra-

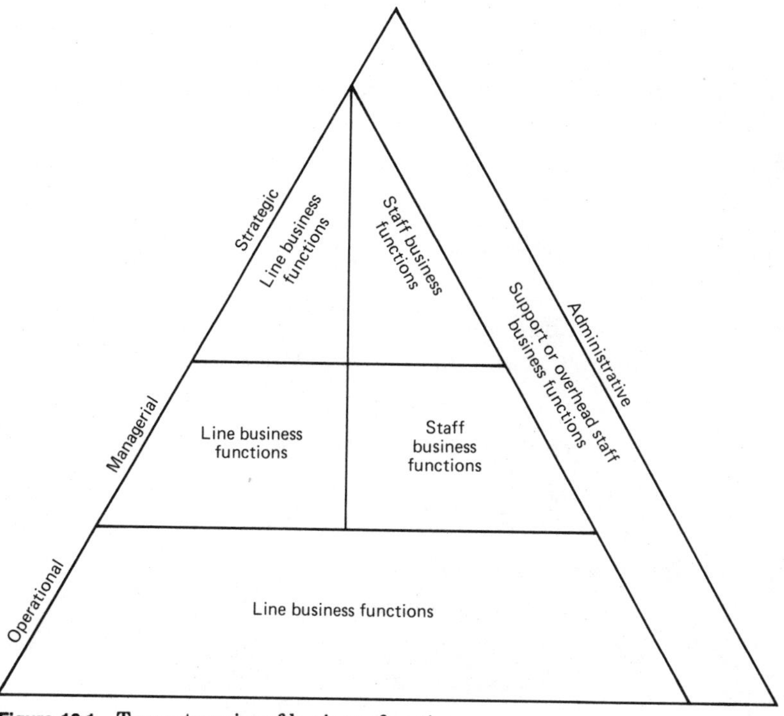

Figure 10.1 Two categories of business functions.

tive category usually contains functions such as personnel, buildings and maintenance, executive management, general accounting, etc. These functions have been termed *staff* functions.

Business Function Analysis

Each user function description should be accompanied by an overview of the business documents or transactions which are input to the user function and which in turn drive that function. Samples of each input form should include form data element descriptions of the information on the form that is currently used by the firm; information that might be of potential use to the firm at a later date should also be included, along with the input descriptions.

The analyst should prepare a functional description for each function addressed. This functional description should be written in user, i.e., business, terms, and should not include any data processing jargon or terminology.

The goal of this phase is to place the user sponsor in perspective within his or her specific business area and within the overall corporate business functional framework.

Again it should be stressed that the language of the document should be in user, and thus in business, terms. The scope of the problem, the length of time the problem has existed, and any attempts which have been made to rectify the problem should also be documented.

Some Questions to Be Asked during Functional Analysis

The business function description document developed as a result of the functional analysis should, at a minimum, answer the following questions.

1. Who are the users? How many people are in the user functional area?
2. What is the average length of employment within the user area?
3. What is the average length of experience in the user area?
4. What is the average rate of personnel head count increase in the user area?
5. What is the average turnover rate in the user area?
6. What do the users do?
7. Whom do the users work for and what do the users' superiors do?
8. What is the business function, or functions, that the user is supposed to perform?
9. What are the users' problems, and why are they problems?
10. Why do the users do what they do?
11. Whom do the users do what they do for?
12. When do the users do what they do?
13. Where do the users do what they do?

Business Function Analysis Documentation

For each input document handled by the user, the analyst should describe the following in business terms.

1. Where do these documents come from?
2. What is their frequency of arrival?
3. What is the average document waiting time and average processing time?

4. What is the average backlog of documents waiting to be handled?
5. What is the document's level of accuracy or completeness?
6. Are there any document processing dependencies? What other information is needed to process the document? Are there any other document dependencies?
7. Are the documents received continuously, or is their arrival cyclic? Are there peaks and valleys?
8. How long are the documents kept?
9. What criteria are used to determine when a document should be disposed of?
10. Are there any document priorities? Are there any significant variations within the general document types?

For each document generated by the user, the analyst should describe the following in business terms.

1. Where do these documents go?
2. What is their frequency of departure?
3. How long does it take to generate the document?
4. What is the average backlog of documents waiting to be generated?
5. What is the document's level of accuracy or completeness?
6. Are there any document dependencies?
7. What is the average length of the document?
8. Are there any significant variations within the general document type?

The documentation produced should also contain, at a minimum,

1. The user organizational chart
2. The placement of the user and his or her organization within the overall organization
3. The nature of the user's business
4. The user's function or job description
5. The user's specific function, or functions
6. The nature of the user's problems, if any
7. The impact of those problems on the user, the user's function, and, if necessary, on the firm
8. Any functional constraints which apply. These constraints should

cover resources, time frames, personnel, business exposure, loss of management control, cost impact, and the costs and benefits of resolving the problem, etc.

Additional Documentation to Include, if Available

In addition the business functional analysis should contain detailed information about

1. The user's charter or functional description and user comments as to how accurate that charter or functional description is
2. Any user-perceived problems with the current functional responsibilities or relationships
3. Any user perceived problems with any of the currently received documents
4. Any user dependencies on functions performed in any other user area
5. Any user management or financial controls

The Need for a Business Function Model

Typically, the business function analysis document alone does not provide a true picture of the scope of the business functions; at best, it provides a limited insight into the interactions of the user's function with the other functional areas of the firm.

Without this insight into the place of the user function within the overall framework of the organization and without an understanding of the relationship of the user function to the other functions of the firm, it is difficult for analysts to assess the accuracy of their business knowledge.

In reality, the various functions of the firm interact in ways which transcend and cross the hierarchic boundaries which are implicit in the table of organization. In many organizations, the table of organization is either restricted to certain title or organizational levels, or is so fragmented as to be meaningless, or worse, hopelessly out of date. (See Figure 10.2 for a manufacturing chart of organization.)

The functional description documentation and the function-to-function relationships developed in this phase should be used to develop a function map, function flow diagram, or functional entity-relationship model.

This business function model (Figure 10.3) places each of the business functions into perspective with every other business function and

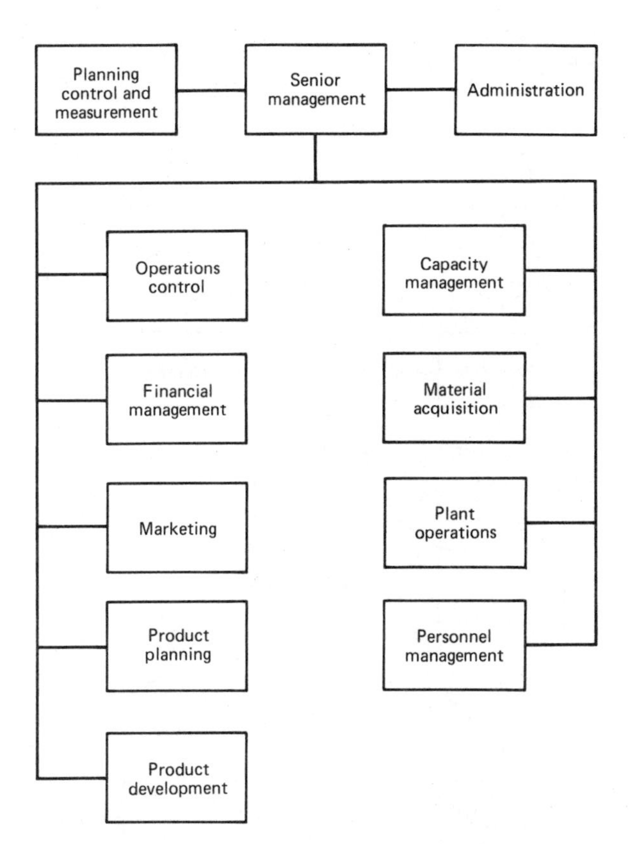

Figure 10.2 Manufacturing chart of organization (to the first level of function).

maps the flow from one business function to another to accomplish the business area function. This function flow diagram, or function map, is also called the *business function model* or *enterprise model*.

The functional model of the firm, or of a specific functional area of the firm, can depict these cross-functional interrelationships and interactions in a much clearer and more meaningful way than can the narrative documentation alone.

The functional model should depict the flow of control, or the flow of materials or information through the corporation, rather than the hierarchic chain of command. In reality corporate functions act and react with one another in much the same way as the components of any complex structure. That is, there is not so much a hierarchic flow as there is a complex set of relationships between quasi-independent units, each of which is responsible for its own functioning, and for receiving and processing material and information and passing it on to others. Sometimes this processing is independent; sometimes it is in

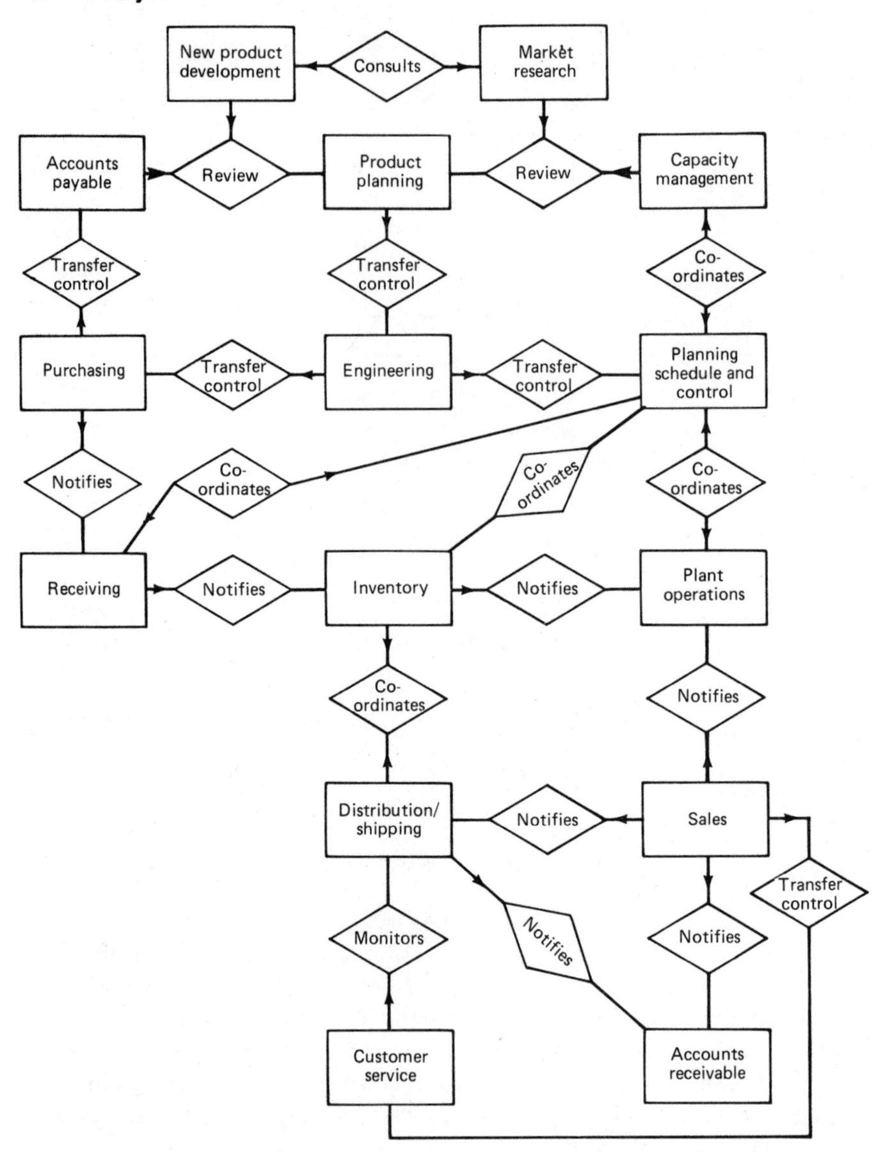

Figure 10.3 Manufacturing functional entity-relationship model (simplified).

concert with other functions. The processing may be linear, or it may be recursive or iterative.

Since functions are concepts, the modeling of functions consists of modeling conceptual things and thus conceptual relationships. One can use the entity-relationship diagrams to depict a functional model.

The concept of a functional entity can be derived by combining the definitions of function and entity.

A definition

A *functional entity* is defined as the conceptual unit which is designed to perform a specific role within the business life of the firm. That conceptual entity performs all the activities and handles all the processing described within the functional description. Each functional entity exists to handle some discrete aspect of the firm's business and thus relates to other functional entities, each of which handles some other discrete aspect of the business. Functional entities normally have the authority to draft, publish, and monitor corporate policies, standards, and procedures related to their specific sphere of responsibilities.

Although the logical starting point of a functional model is the corporate organizational chart, there are some problems when it is used to identify the business functions.

1. Functions are not properly placed within the organization.
2. Functions are not properly named in an organization.
3. Functions either do not appear on the chart of organization or are not identified by the user.
4. Functions are not readily identifiable, except in a graphic model.
5. Some functions are in reality multiple functions combined under one name.
6. The interaction of the various functions cannot be appreciated except when placed in the overall context of the firm.

Since functions are managerial concepts, the relationships between functional entities are by definition either managerial, consultative, or simply transfer-of-control relationships.

1. Managerial relationships are those where one entity performs in a planning, controlling, monitoring, organizing, coordinating, or reviewing capacity over another.
2. Consultative relationships are those where two independent functions work together to perform an activity or achieve a common result. A consultative relationship may also be one of information transfer (informative) or advisory.
3. Transfer-of-control relationships are those where responsibility for the completion or continuation of a particular task passes to a new

function, or where a task has been completed and that completion triggers the beginning of new tasks by new functions. A transfer-of-control relationship may be implemented by the passing of some physical object, such as a file, document, person, or thing from one functional area to another, or it may be implemented by simple notification, where one function notifies another that a state or condition has changed and that the second function may now initiate its activities.

In the case of support functions, there may be an additional relationship of a request for service. That is, one function may request the services of another and that request itself initiates a sequence of activities. In some cases more than two functions may participate in a relationship or may contribute the results of their activities to still another function.

The relationship between two functions should be expressible in terms of simple sentences. Each sentence should have a subject (function one), a verb (the relationship), and an object (function two). To illustrate: Purchasing notifies receiving to expect an incoming shipment from a vendor (Figure 10.4).

The Business Life Cycle Matrix

A business life cycle represents the flow of control and interaction through the business. The relative positioning of each functional box is determined by its activity during the life of a single product of the firm, or other unifying aspect or thread.

The determination of this thread is one of the more difficult aspects of functional analysis. One primary source in determining this thread might be the corporate mission statement or charter. Another might be statements by senior management in the annual report as to the firm's main business, or lines of business.

Where the firm has multiple lines of business, the analysts should expect multiple threads to appear, unless the lines of business represent a vertical or horizontal integration of the overall business. Some functions are active only once during the life of a product, or on a thread, while others are active on a continual basis. For the purpose of

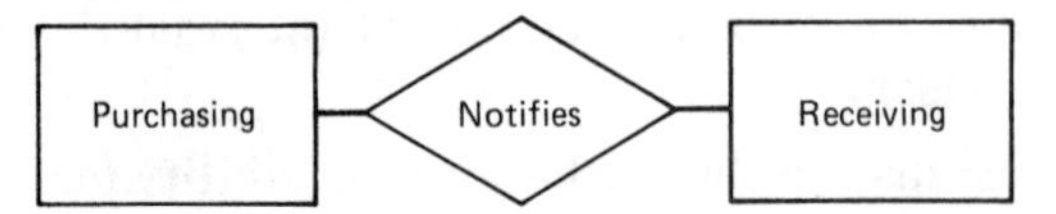

Figure 10.4 Relationship between purchasing and receiving.

the model, one should assume that each function is active for the duration but has a specific activation point.

Developing the Functional Model

Developing a functional model requires the following steps.

1. Working from a corporate table of organization, the user interviews, and the functional area descriptions, the analyst is able to identify the individual functions of the firm, develop detailed descriptions of each function, the role of those functions within the firm, and the functions which interact with the subject function. Since each function plays a unique role, it must interact with other functions which play other roles.
2. Determine those interactions by examining those functions to which the subject function has a direct or immediate relationship.
3. Make each functional box in the table of organization into a functional entity. Care must be exercised at this point to ensure that each of these functional entities represents a single, unique function, not a combination of functions. All subfunctions should be broken out into separate boxes for clarity of presentation.
4. Arrange these functional boxes with the names of the functions within each box in a rough sequence corresponding to the business life cycle being depicted.
5. Draw the relationships between each set of functions as they relate to each other during the particular life cycle being depicted.
6. Add the name of each relationship to each relationship symbol, and note the flow of the lines connecting the relationship symbols to the functional entities. Unlike the typical entity-relationship models, the relationships will not normally be bidirectional. That is, the flow will normally pass from function 1 to function 2, but not necessarily from 2 back to 1.

When completed, the functional model should depict the flow of control, material, or information, or all three, along the life cycle. Because the firm may have many products, services, information, controls, or other logical threads, it may be necessary to produce multiple models, each depicting a particular thread or flow. For instance, the accounting thread will probably be different from the material management thread. The various lines of business threads may themselves be different from the administrative thread. In decentralized organizations there might even be multiple versions of each of the basic threads, each corresponding to the management techniques in use by each decentralized unit. These decentralized threads might in turn be different from the parent company threads.

For these reasons it may be necessary to produce multiple models, or to clearly state at the outset which part of the firm is being modeled. If multiple models are produced, they may intersect or overlap at various points or they may be completely discrete. Overlap and intersection points, if they exist, should be examined carefully to determine if both models agree at those points.

Chapter

11

Process Analysis

CHAPTER SYNOPSIS

The business process and activity analysis step is a more in depth analysis of the information and business functions identified and developed in the preceding steps. A business function is an end-to-end role or reason for existence. These individual functions, however, are composed of a series of processes or individual activities which individually and in combination perform the work of that function. A function is thus a group of related processes or activities all performed to achieve a predetermined goal. Those discrete processes or activities are in turn composed of discrete tasks.

This chapter discusses the concept of a business process and provides some insight into their identification and documentation. It includes extensive lists of questions for the analyst to ask when interviewing at the process and activity level.

What Is a Business Process?

A definition

A *process* is a sequence of related activities, or may be a sequence of related tasks which make up an activity. These activities or tasks are usually interdependent, and there is a well-defined flow from one activity to another or from one task to another.

A definition

An *activity* is a set of tasks which are organized and proceduralized to accomplish a specific goal. The distinction between a subfunction and an activity is as much a matter of interpretation as it is a matter of scope.

Generally, an activity is discrete and part of an overall function. It is highly structured and task oriented as opposed to control or management oriented. Generally functions are managed and activities are performed, although this distinction is far from definitive.

Activities are data driven in that they are triggered by transactions or requests for data. Activities are the active portion of functions and tend to be repetitive and formalized.

Business Process Analysis

The business process and activity analysis step is a more in depth analysis of the information and business functions identified and developed in the preceding steps.

A business function is an end-to-end role or reason for existence. These individual functions, however, are composed of a series of processes or individual activities which individually and in combination perform the work of that function. A function is thus a group of related processes or activities all performed to achieve a predetermined goal. Those discrete processes or activities are in turn composed of discrete tasks.

A business process is a set of activities or tasks which are performed in sequence or in parallel to accomplish a specific goal. The process may be manual or automated and may comprise one or more activities or tasks.

For each function identified at the preceding step, the analyst must identify those processes or activities which are performed to support that function. They may be performed by one or more persons in one or more areas of the firm.

A major portion of the effort in the analysis phase is devoted to this analysis of user processes and activities. This analysis describes and defines each user process and activity to determine if, and under what conditions, they may be automated. As with the functional analysis these descriptions in and of themselves do not provide adequate insight into the flow, complexity, or interdependence of the user processing activities. All too often each user process is treated as if it stood alone and was complete in and of itself. This individual treatment of processes does not capture the processing flow, nor does it necessarily detect related processes which are external to the immediate user's function.

The task of the analyst is to identify and describe in detail

1. All the areas where those related processes or activities are accomplished

2. What those processes or activities are
3. How they are performed
4. Why they are performed
5. How each process or activity relates to the other processes and activities of the function and to other processes and activities within other functions

Each process or activity must be identified, described, and the following information gathered and documented.

1. The business transactions which trigger this process or activity.
2. For each transaction, a description of its source document, its timing, frequency, and volume, any variants, all exception conditions, and any editing which is performed.
3. If only part of the data resident on the transaction source document is necessary, a statement as to why the remainder is unused. Any additional sources of the same data should be identified, as well as the rules for validation and verification.
4. For each transaction, describe in detail the processing which is initiated as a result of that transaction's arrival. Document whether that processing is immediate, deferred, or conditional. If the processing of that transaction is dependent upon prior processing or prior data availability, it should also be indicated.
5. Any applicable manual procedures which apply to the processing should be identified, described, and, where applicable, included.
6. Any data saved or filed between processing steps should be identified and the method of saving or filing the data described. Data saved on the originating transaction document should be included in this description.
7. Any reports generated as a result of the processing should be documented, and samples of each unique page of each report should be included in the documentation. Any controls should be documented, and any auditing procedures should be described.
8. Any control checks, control points, and decision points should be clearly identified and described.
9. All procedures for error detection and correction should be clearly documented, and the actions taken as a result of error detection should be included.

Some Questions to Be Asked during Process Analysis

The business process description document developed as a result of the process analysis should, at a minimum, answer the following questions.

1. Who are the users? How many people are in the user processing area?
2. What is the average length of employment within the user area?
3. What is the average length of experience in the user area?
4. What is the average rate of increase in personnel head count in the user area?
5. What is the average turnover rate in the user area?
6. What do the users do?
7. Whom do the users work for, and what do the users' superiors do?
8. What is the business process or activity, that the user is supposed to perform?
9. What are the user's problems, and why are they problems?
10. Why do the users do what they do?
11. Whom do the users do what they do for?
12. When do the users do what they do?
13. Where do the users do what they do?

Where possible, the documentation should answer the following questions.

1. Who performs the processes or activities?
2. What are the actions performed by that person or persons?
3. When are these actions performed?
4. Why are these actions performed?
5. Why are these actions being performed at this time?
6. What is the impact on the firm if these actions are not performed?
7. What is the impact if these actions are performed incorrectly?
8. What is the frequency of these transactions?
9. What is their rate of arrival?
10. Are there peak periods?
11. What is the average number of transactions processed per day, week, or other applicable time frame?

12. What is the maximum expected number of transactions?
13. Are there any problems which consistently arise with the present transaction documents?
14. Are there any data which could be added to the document which would ease the processing burden?

Business Process Analysis Documentation

For each input document handled by the user while performing the process or activity, the analyst should describe in business terms

1. Where do these documents come from?
2. What is their frequency of arrival?
3. What is the average document waiting time and the average processing time?
4. What is the average backlog of documents waiting to be handled?
5. What is the document's level of accuracy or completeness?
6. Are there any document processing dependencies? What other information is needed to process the document? Are there any other document dependencies?
7. Are the documents received on a continuous even basis, or is their arrival cyclic? Are there peaks and valleys?
8. How long are the documents kept?
9. What criteria determine when a document is disposed of?
10. Are there any document priorities? Are there any significant variations within the general document types?

For each document generated by the user, the analyst should describe in business terms

1. Where do these documents go?
2. What is their frequency of departure?
3. How long does it take to generate the document?
4. What is the average backlog of documents waiting to be generated?
5. What is the document's level of accuracy or completeness?
6. Are there any document dependencies?
7. What is the average length of the documents?
8. Are there any significant variations within the general document type?

The documentation produced should also contain, at a minimum,

1. The user's job description.
2. The nature of the user's problems, if any.
3. The impact of those problems on the user, the user's processing, and, if necessary, on the firm.
4. Processing constraints which apply. These constraints should cover resources, time frames, personnel, business exposure, loss of management control, cost impact, and the cost versus benefits of having the problem resolved, etc.
5. User-perceived problems with the current processing responsibilities or relationships.
6. User-perceived problems with any of the currently received documents.
7. User dependencies on processes performed in any other user area.
8. User management or financial controls should also be well documented.

Additional Documentation to Be Included, if Available

In addition the business process analysis should contain detailed information about

1. The user's charter or job description, and user comments as to the accuracy of the charter or job description.
2. Any standards, policies, or procedures which govern or apply to the user's processing activities.
3. Any user manuals which have been produced by the user or for the user by others.
4. Samples of all documents handled by the user area. These samples should include both blank and completed documents and should include as many exception cases as possible.
5. Samples of all reports, charts, graphs, and other output produced by the user area. These reports should be accompanied, where possible, by a complete description of each section or entry on the document, report, graph, chart, etc.

The Need for Business Process Models

Just as the business functional analysis benefits from the development of a business function model, so too the business process analysis benefits from the development of business process models, one for each major, or significant, business process. These models depict the interactions of the various user tasks and activities during the performance of the processing and the interaction of the user's processing with other processing areas of the firm.

Without this graphic representation of user processing within the overall framework of the function, and perhaps within the organization as well, and without an understanding of the relationship of the user process with the other processes of the firm, it is difficult for analysts to assess the accuracy of their business knowledge and their understanding of the information gathered from the user during the interview process and from their own observations.

Many processing flows interact in ways which transcend and cross the hierarchic boundaries which are implicit in the table of organization. The process models, both individually and in composite form, help the analyst understand the overall flow of processing and of data and information within the firm.

The process description documentation and the process-to-process relationships developed in this phase will be used to develop a process map, process flow diagram, or process entity-relationship model.

These business process models place each of the business processes into perspective with every other business process and map the flow from one business process to another to accomplish the business area function.

As with the business function model, the process model of the function can depict these interrelationships and interactions in a much clearer and more meaningful way than can the narrative alone.

Business process models may be produced in a variety of ways and for a variety of reasons. They may model the flow of material or information into and out of processing stations, or they may model the flow of a particular item of material or information through the entire firm, showing its various uses and transformations. The models may attempt to show the sequence of processes without regard for the inputs and outputs. Some models indicate conditional flows and others ignore conditional logic.

For this reason, as with the functional models, it may be necessary not only to develop multiple models but to use many different modeling techniques to portray all the relevant information. Entity-relationship models may be drawn, for example, to depict the sequence of processes relevant to a particular input document or type of material.

Data flow diagrams, on the other hand, might be used to depict all of the inputs and outputs of a series of related processes and the ways in which all the inputs and outputs affect each other.

A hierarchic process diagram would depict the processing dependencies and sequence without regard to the particular data or materials which drive, or result from, the processes themselves.

Flowchart diagrams may be used to depict not only the processes but also the types of data storage or transfer media involved and the logical decision points which govern the various processing legs.

Business Process Entities versus Processes

Business processes are the activities performed by a business functional unit. Because the processing units and the processes performed by those units usually bear the same designation, any model produced should clearly indicate whether the processing stations or the processing activities themselves are being modeled. The need for this distinction becomes apparent when one looks at the differences between the two. The concept of a processing station or a processing entity is derived by combining the definitions of a process and an entity.

A definition

A *process entity* is defined as the conceptual unit which is designated to perform one or more specific processing tasks upon a business transaction or upon business materials. That conceptual entity performs all the activities and tasks required by a single processing step. Each process entity exists to handle some discrete aspect of a business transaction and thus relates to the other process entities, each of which handles some other discrete aspect of that transaction.

In many businesses the same discrete set of process activities may be performed in many different physical units (a logical flow). By the same token, a physical unit may perform a number of different discrete processes (a physical flow). These processes may or may not be related to each other through a particular transaction or thread.

It thus becomes necessary to state clearly whether we are modeling the flow of processes logically between units or physically within a unit itself, or whether we are modeling the processing actions themselves. The logical flow follows the transaction through the firm; here the transaction remains constant and the processing areas change. The physical flow models the inputs and outputs of a particular unit; here the unit remains constant and the flows change. The logical flow is the more difficult of these models, since, as stated previously, process entities interact in ways which transcend and cross the hierarchic boundaries which are implicit in the table of organization. A process model can depict these interprocess relationships and interactions in a much clearer and more meaningful way than can a straight narrative.

When we dealt with functions, we dealt with their roles or, in other words, what function was performed and when. When we examine process entities we are again dealing with the roles or activities of the process entities within the functions. These process entities are of necessity subservient to the functional entities and the functional relationships which have been previously defined. A functional entity should not employ a process entity which is not related to the functional role itself; however, it is possible to find process entities which perform work for multiple functional entities. In this case, the processes should be consistent with each of those functions. Personnel record keeping activities would be an example of such a cross-functional process.

In the functional model we took care to ensure that each functional entity represented a single functional role. Within a process model we must take care to ensure that each process entity represents a single process. Each unique process should be triggered or initiated by a single transaction type or class of transaction. Each process may require other material or data to complete its tasks, but there should be a primary trigger transaction.

Process Entity-Relationship Models

A process entity-relationship model is not concerned with the data that flows between the process entities but rather with the direction of the flow, source, and destination of the flow and the sequence of the stations along the flow. Here we are not concerned with the processing that is performed or even how it is performed, but rather who performs the processing and when.

These are not necessarily transaction flows in that they follow a particular transaction through the firm, but rather they depict the sequence of actions which result from a particular transaction. Just as there may be many functional threads through the firm, so too there may be many separate processing threads; however, no transaction should generate multiple threads, although a single thread may have multiple legs (see Figures 11.1, 11.2, and 11.3).

While these multiple threads may cross, join, or run parallel, each is distinct and thus each should be modeled separately. The usual method for producing process models is to select each distinct primary document, in turn, and note the sequence of processing steps which are initiated by it.

The sequence will end with a final report being produced, the information contained in the transaction being stored in the files of the firm, or the information being of no further interest to the firm.

Where the sequence ends in a filing step, it should be noted whether this is an intermediate file, waiting further triggers (such as an invoice awaiting payment or a purchase request awaiting material re-

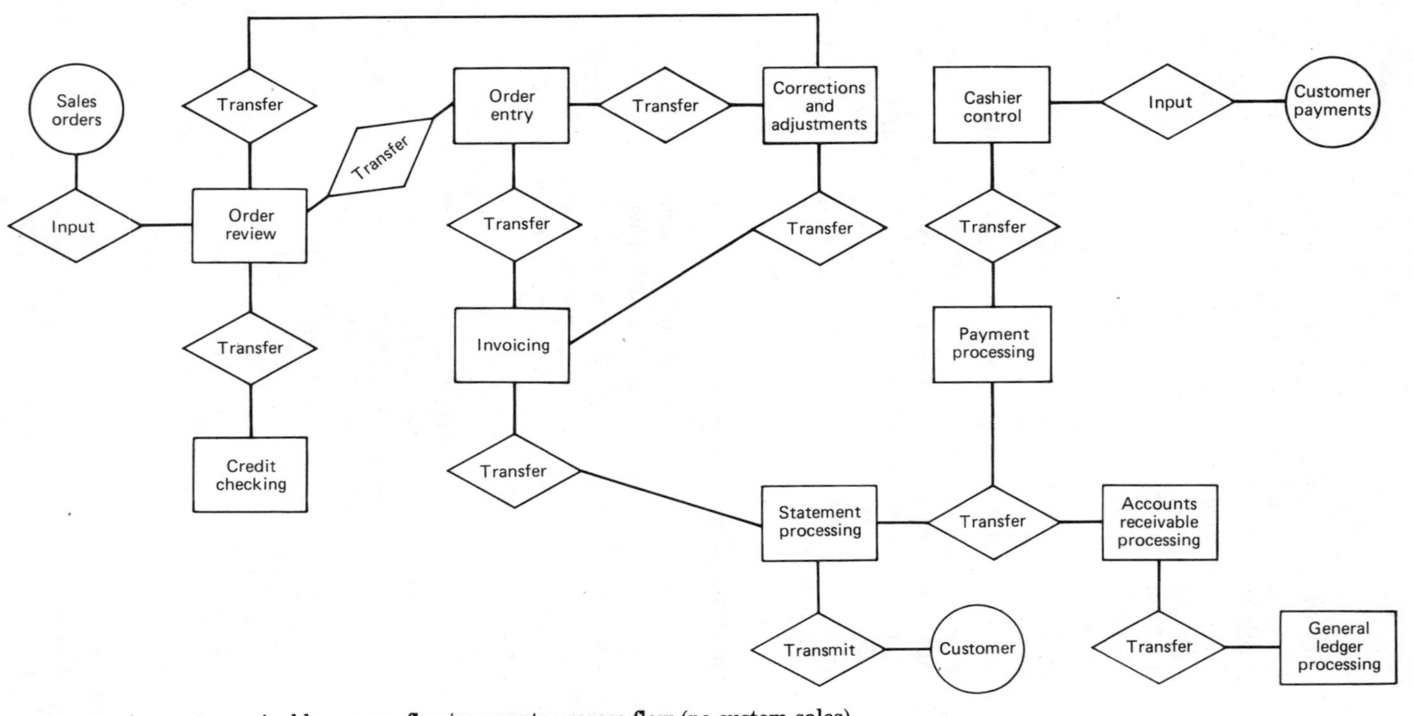

Figure 11.1 Accounts receivable process flow/payments process flow (no custom sales).

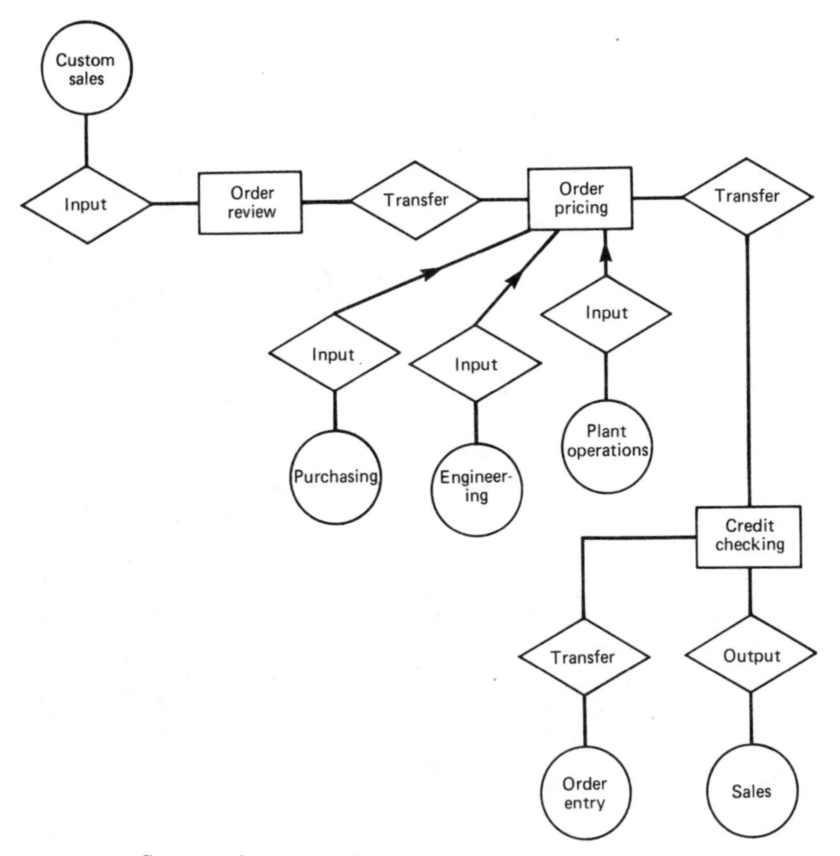

Figure 11.2 Custom sales pricing flow.

ceipt) or a final or archive file (such as paid invoices, filled orders, or shipped material).

Guidelines for Developing a Process Model

The following are some guidelines to be used in developing a process model.

1. Clearly identify the subject of the particular model.
2. Clearly identify the start and end points.
3. Clearly identify the name of each process or process entity in the model.
4. Identify the direction of the flow, any relationships between pro-

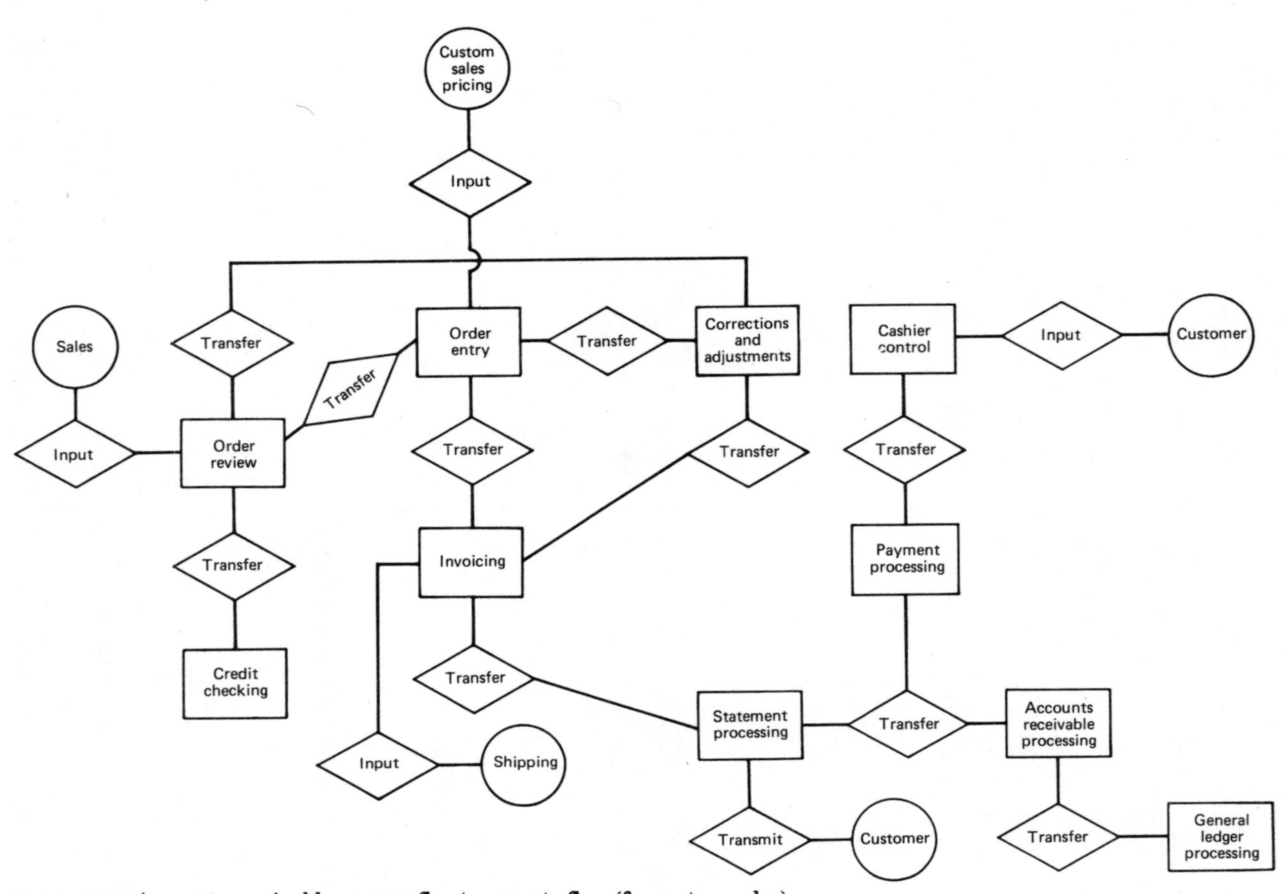

Figure 11.3 Accounts receivable process flow/payments flow (for custom sales).

cesses, and of any material or information entering or leaving each process step or entity.

5. Identify the time sequences or time frames involved.
6. Identify the source of any off-model inputs, and the destination of any off-model outputs.
7. Clearly identify the thread of the model and any points where the activities of this thread may initiate other threads.

It should be understood here that we are looking at a sequence of steps, not necessarily at the flow of the transaction. The transaction is only the trigger for the first step.

Chapter

12

Activity or Task Analysis

CHAPTER SYNOPSIS

A task is the lowest unit of work within the scope of the analysis. Task analysis is the final level of system decomposition. These tasks may be performed in a manual, automated, or semiautomated manner. Each task has one or more inputs, some processing, and some outputs. The analyst must carefully examine the individual tasks being performed by the user to determine how, when, why, and under what conditions the task is to be performed.

This chapter discusses the concept of a business task and provides some insight into its identification and documentation. It includes extensive lists of questions for the analyst to ask when interviewing at the task level.

What Is a Task?

A definition

A *task* is the lowest unit of discrete work which can be identified. An activity may be composed of many tasks. Tasks are highly repetitive, highly formalized, and rigidly defined.

Business Task Analysis

A task is the lowest unit of work within the scope of the analysis. Task analysis is the final level of system decomposition. These tasks may be performed in a manual, automated, or semiautomated manner. Each task has one or more inputs, some processing, and some outputs. The analyst must carefully examine the individual tasks being performed by the user to determine how, when, why, and under what conditions the task is to be performed.

Although it is not the analyst's goal to perform the user's work, the analyst must know as much as, if not more than, any individual user about the actual mechanics and rationale of that work. Many analysts concentrate on the mechanical aspects of the task performed by the user and thus develop automated systems which replicate those tasks within the machines. While understanding the task mechanics is important and automation of those tasks is a goal, it is also vital to understand why those tasks are being performed and what the effect of that task performance is on other tasks within the immediate user area and its larger impact on the firm.

No task exists or should exist in a vacuum. Tasks are performed for a business reason, and the specific manner in which they are performed also has a business reason. The determination of those business reasons is as much a part of the analysis process as is the determination of the task mechanics. Task analysis is not usually performed as a specific step within the overall analysis phase, but it is usually combined with the process analysis activities. It is presented separately for clarity and to highlight task-specific questions.

The analyst must remain aware of a basic fact of business systems: Tasks are added to business systems as the need arises, usually with no overall plan in mind. The need is recognized, analyzed (it is hoped), a procedure written (although not always), and someone is (always) assigned to the new job. These tasks, or jobs, "grow like Topsy." In many instances what were understood to be one-time, or limited-term, tasks become institutionalized. These jobs may be added to existing process flows, but, more likely, new process flows are created and "hooks" added to facilitate communications with the existing structures.

Over time, these makeshift procedural structures also become institutionalized. Each of these new jobs and procedural structures evolves over time and takes on new characteristics which may never have been conceived of initially. Rarely, if ever, does anyone look to systematize these diverse tasks. There is a basic law of organizations which postulates that work expands to fill the available time.

Since these new tasks are rarely, if ever, added according to any plan and are in fact created to satisfy an immediate need, they have a tendency to contain redundant activities, overlap existing activities, and in some cases they may even be in conflict with existing activities. Moreover, because they were created in haste, they may be misplaced within the organization as well. That is to say, they may be performing tasks which may rightfully be the responsibility of some other functional area. Often, these new activities or tasks were created and placed within an area because it was easier to do so than to get the rightful area to do the work, or because the rightful area was too busy to do the work and the chosen area had free resources.

Each task should be examined in this light. Each form emanating from each business task should be examined with the understanding that no one may know where it originated or why. Many business forms and the tasks which either process or create them may be redundant, or worse unnecessary. The same reasoning holds true for business procedures.

Some Questions to Be Asked during Task Analysis

1. What is being done?
2. How is it being done?
3. Why is it being done?
4. Should it be done at all?
5. If it needs doing, is there a better way to do it?
6. Should the user be doing it or should someone else?
7. What can be done to make the task simpler?
8. Can the tasks be rearranged to make the activity or process more streamlined?
9. Does the user have all the resources and all the information needed to perform the task?
10. Can the user complete the task in one operation, or is the user dependent upon something or someone else to complete the task?
11. Is this user the only one performing this task, or is it being performed in an identical or highly similar manner by others within or outside of the user's immediate area?
12. Are there specific standards or procedures governing this task? Are they well documented? When were they last revised or reviewed? Are all user personnel assigned to the task familiar with these standards and procedures?
13. How much discretion is allowed in the performance of the task?
14. How often are exception conditions encountered, and how are they handled? How long does it normally take to resolve these exceptions?

15. Are the personnel informed on a routine basis as to the resolution of these exceptions?
16. Can the processing be revised, or can something else be done to eliminate or minimize these exceptions?
17. Is the task relatively straightforward, or are there decisions which have to be made during the processing? How may decisions need to be made? Can the decisions be answered objectively (with a yes or no or with a number), or must they be answered subjectively?
18. Is the sequence of task steps straight line or does it have loops or multiple legs?
19. On average, how long does it take to accomplish a single iteration of the task (e.g., process one document)?
20. Are there performance standards which govern the task? If so, what are they? Are they regularly met? Does the worker have difficulty meeting the standard? What is the impact of not meeting the standard?
21. Do the workers require any special education or training to perform the task? What is that special education or training? How long does the training period last?
22. How often are the training materials reviewed, revised, or otherwise updated?

Task Modeling

Because of the nature of tasks as opposed to processes and because when analyzing tasks we are interested in how the task is accomplished, task models are usually produced in flowchart form.

The flowcharts should begin with the task trigger (a document receipt, a communication of some sort, or the movement of some material into the task area). Each manual *and mental* step should be shown along with each decision point. Each calculation and each test should be clearly labeled and documented.

For each decision point, the decision question itself should be placed in the decision point symbol, each possible result of the decision (yes or no, present or absent, greater than, less than or equal) should be noted, and the steps which result from each possible result should be shown.

For clarity, the flow should begin in the center of the page, and each new decision action leg should be placed on a separate page as well. Any nonstep information should be noted in the margins of the page along with form and reference notations as well. Calculations for steps requiring them should be noted in the margins as well.

Flowcharts should be annotated with the name of the task being depicted, the date the chart was produced, the analyst's name, and the name of the user source and the user reviewer. Each chart and each page of each chart should be numbered, and provisions should be made for chart revision information.

Chapter 13

Data Analysis

CHAPTER SYNOPSIS

Data analysis is that process which identifies, element by element, data requirements of a functional area. Each data element is defined from a business sense, its ownership is identified, users of that data are identified, and its sources are identified. These data elements are grouped into records, and a data structure is created which indicates the data dependencies.

Data analysis focuses on two aspects, data currently used by the user and data that will or should be needed in the future by the user. Current data is analyzed further to determine if it is being collected from the most accurate source, at the right time, and at the right level of detail. Current data analysis also tries to determine whether the correct business definitions of that data are being employed and whether all users of the same data define it and view it in the same manner.

This chapter discusses the need for data analysis and the various techniques for accomplishing data analysis and producing the documentation which should be assembled as a result of it.

What Is Data Analysis?

A definition

Data analysis is that process which identifies, element by element, data requirements of a functional area. Each data element is defined from a business sense, its ownership is identified, users of that data are identified, and its sources are identified. These data elements are grouped into records, and a data structure is created which indicates the data dependencies.

Data analysis focuses on two aspects: data currently used by the user and data that will or should be needed in the future by the user. Current data is analyzed further to determine if it is being collected from the most accurate source, at the right time, and at the right level of detail. Current data analysis also tries to determine whether the correct business definitions of that data are being employed and whether all users of the same data define it and view it in the same manner.

What is data event analysis?

A data event is something which happens within the business environment which the company needs to know about and which must be recorded in the company memory, that is, the company files. A data event may be externally or internally generated and may occur through some action being taken or merely as a result of the passage of time.

The occurrence of data events recorded in some manner. Data event analysis determines what information must be recorded such that the event can be recalled and acted upon. It must also determine how that event became known to the company; that is, what triggered the company awareness of the event?

What is transaction analysis?

Transaction analysis is coupled with data event analysis. Transaction analysis looks at the data carriers which move data and information around the firm. Some of these transactions may be externally generated and some are internally generated.

What is document and forms analysis?

Document and forms analysis is a subset of data event and transaction analysis, and looks at the forms and documents which carry data through the firm. Its aim is to determine all the causes of data and whether the firm is saving and using all the data from those forms and documents in the most efficient manner. Document and forms analysis also looks to see whether the forms and documents are designed well; that is, are the data correctly identified; is there sufficient room on the form for the data to be entered; are the data clustered on the form properly; are there enough copies of the form; how, when, and where are they filed? Are the forms retained for the proper length of time; are they secured properly; are they filed and indexed properly; and can they be retrieved in a reasonable length of time?

What is report analysis?

Report analysis concentrates on the outputs of data processing, regardless of whether that processing was automated or manual. Reports provide users with information deemed important to their activities. The analyst must determine whether all the reports received by the user are necessary and whether they are accurate, timely, or complete. Report content and report documentation must also be analyzed to determine if the user's understanding of the report's contents coincides with what the report actually presents. The analyst also must determine whether the user is receiving the report in enough detail, or too much detail, and whether it provides accurate or complete totals.

Data Analysis Questions

Although we have presented data analysis as a separate section, the questions pertaining to data have appeared in the chapters on function, process, and task analysis. The documentation of data has been discussed in the sections on the dictionary and in most of the preceding sections as well.

Part

3

Validation

MURPHY'S LAWS

The most important piece of information in any plan, or document, stands the greatest chance of being left out.

The most important person, piece of material, or piece of information is the one that's missing.

Things that cannot possibly be done in the wrong order will be.

In any collection of data, the part that is most obviously correct and does not need checking—is the mistake.

*The error detection and correction capabilities of any system will serve as the key to understanding the type of error which the system can***not** *handle.*

Undetectable errors are infinite in variety, in contrast to detectable errors, which, by definition, are limited.

Nothing is ever as simple as it seems.

Chapter

14

Validating the Analytical Results

CHAPTER SYNOPSIS

The process of analysis validation is as important as the original analysis. There are numerous methods available to analysts to validate their work, but not all validation techniques work equally well for all types of analysis.

This chapter describes some of the most commonly used techniques, explains how they can be used effectively, and provides a discussion as to the conditions under which they should be used, and the advantages and disadvantages of each.

Validating the Analysis

Having completed the analysis process and having documented the results, the analyst's last and most critical steps are directed toward the validation of the work. The validation process is performed to ensure that

1. All parties agree that the conditions as presented in the documentation accurately represent the environment.
2. The documents generated contain statements that are complete, accurate, and unambiguous.
3. The conclusions presented are supportable by the facts as presented.
4. The recommendations address the stated problems and are in accord with the needs of the users.

The analytical results not only represent the analysts' understanding of the current environment and their diagnosis of the problems inherent in that environment, but also their understanding of the user's unfulfilled current requirements and projected needs for the foreseeable future. A combination of these four aspects of analysis will be used to devise a design for future implementation. Thus it is imperative that the analysis be as correct and as complete as possible.

Just as there are multiple techniques available during the analysis process itself, so too there are multiple techniques available to validate the analysis. The aim of the validation process is to ensure that all the pieces have been identified, understood individually and in context, and described properly and completely.

A system is a complex whole. Old systems are not only complex, but in many cases they are a patchwork of processes, procedures, and tasks which were assembled over time and which may no longer fit together into a coherent whole. Many times needs arose which required makeshift procedures to solve a particular problem. Over time these procedures become institutionalized. Individually they may make sense and may even work; however, in the larger context of the organization, they are inaccurate, incomplete, and confusing.

Most organizations are faced with many systems which are so old and so patched, incomplete, complex, and undocumented that no one fully understands all of the intricacies and problems inherent in them, much less has a complete overview. The representation of the environment as portrayed by the analyst may be the first time that any user sees the entirety of the functional operations. If the environment is particularly large or complex, it could take both user and analyst almost as long to validate the analysis as it did to generate it, although this is probably extreme. The validation of the products of the analysis phase must address the two aspects of the environment: data and the processing of data.

Analysis seeks to decompose a complex whole into its component parts. For data it seeks

- To trace the data flows into, from, and through the organization
- To identify and describe all current and proposed data inputs, and to determine the organization's need for the various component informational elements of those inputs
- To identify and describe all current and proposed data outputs and to determine the value, completeness, and accuracy of those outputs and their relevance to the intended recipients
- To identify all current and proposed data stores (ongoing files), and to determine the value, completeness, and accuracy of those data stores and their relevance for their intended owners

For processes it seeks

- To trace the processing flows and their component tasks, individually, in relation to each other, and in the context of the user function and the overall functions of the organization
- To identify and describe all current and proposed data inputs which trigger those processes and the outputs which result from those processes, and to determine the organization's need for the process
- To identify and describe all current and proposed results of the individual processes, the completeness and accuracy of the processing, and the relevance of that processing to their intended recipients
- To place the process within the larger context of the functions of the firm

Validation seeks to ensure that the goals of analysis have been met and that the results of each of the three component parts of the analysis—current environment, problem identification, and future environment proposal—are complete and accurate, solve the user's problems, and reflect the expressed needs of the user and of the business. It also attempts to ensure that for each component, the identified component parts re-create or are consistent with the whole.

To use an analogy, the analysis process is similar to taking a broken appliance apart, repairing the defective part, and putting the appliance back together again. It is easy to take the appliance apart, somewhat more difficult to isolate the defective part and repair it, and most difficult to put all the pieces back together so that the appliance works. The latter is especially true when the schematic for the appliance is missing, incomplete, or, worse, inaccurate.

The documentation created as a result of the analysis is similar to a schematic created as the appliance is being disassembled; the validation process is similar to trying to put the appliance back together using the schematic you created. In many respects the validation process is like the testing processes during implementation. The system may appear to be complete and may appear to have addressed all possible conditions, but that can only be assured by running exhaustive tests. These tests are usually run first at the unit level, then at the procedure level, and finally at the system level.

As with implementation testing, one must run through all possible combinations of logical pathing and must run all possible transaction types through the system to ensure that the implementation handles them properly. Test data should not be created by the people who wrote the code; likewise analysis validation should run through all possible logic paths, test all possible transactions, and be conducted by persons *not* associated with the actual analysis. The techniques for validation of the products of analysis seek to ensure that the verifier sees what *is* there and not what *should* be there.

By its very nature, systems analysis works to identify, define, and describe the various component pieces of the system. Each activity and each investigation seeks to identify and describe a specific piece. The piece may be macro or micro, but it is nonetheless a piece. Although it is usually necessary to create overview models, these overview models, at the enterprise and functional levels, seek only to create a framework or guidelines for the meat of the analysis, which is focused on the operational tasks. It is the detail at the operational levels which can be validated. The validation process of both data and process work at this level. Each activity, each output, and each transaction identified at the lowest levels must be traced from its end point to its highest level of aggregation or to its point of origination.

Many of the techniques which are employed in the analysis phases themselves may be used to validate that analysis. The primary differences between using a technique during original analysis and using it during validation are in what they are applied to and what the analyst is attempting to achieve.

In the analysis, the analyst is gathering facts, seeking to put together a picture of the current environment, and diagnosing that environment to determine any flaws in its structure or points where improvements may be made. By using this information, the analyst can create a picture of the environment as it could and should be in the future. At the outset, both the environment and the format and content of the pictures are vague at best and unknown at worst. If the analysis and the diagnoses are correct, the proposed environment will satisfy the user's needs both immediately and in the long run.

During validation, the analyst begins with an understanding of the environment and the pictures or models that have been constructed. The aim here, however, is to determine

- Whether the analyst's understanding of the environment is complete and correct
- Whether the depictions of the current environment matches what is actually there and the user's understanding of the environment
- Whether the analyst's diagnoses correspond with the user's own perception of the problems and areas for improvement
- Whether the proposed future environment will satisfy the user's perceptions of his or her immediate and long-term needs

It must be understood that the analysis, diagnoses, and proposals represent a combination of both fact and opinion. They are also heavily subjective. They are based upon interview, observation, and perception. The purpose of validation is to assure that perception and subjectivity have not distorted the facts.

The generation of diagrammatic models at the functional, process, and data levels greatly facilitates the process of validation. Where these models have been drawn from the analytical information and where they are supplemented by detailed narratives, the validation process may be reduced to two stages.

Stage 1

The diagrams are cross-referenced to the narratives to ensure that

1. Each says the same thing.
2. Each figure on the diagram has a corresponding narrative, and vice versa.
3. The diagrams contain no unterminated flows; there are no unconnected figures or ambiguous connections; all figures and all connections are clearly and completely labeled and cross-referenced to the accompanying narratives.
4. The diagrams are consistent within themselves, that is, data diagrams contain only data, process diagrams contain only processes, and function models contain only functions.
5. Each diagram is clearly labeled and a legend has been provided which identifies the meaning of each symbol used.
6. When the complexity of the user environment is such that the models must be segmented into many parts, each part is consistently labeled and titled, the legends are clear, connectors between the parts are consistent in their forward and backward references, and names of figures which appear in different parts are consistent.

Stage 2

Cross-referencing across the models ensures that

1. Processes are referenced back to their owner functions, and functions reference their component processes.
2. Any relationships identified between data entities have a corresponding process which captures and maintains them.
3. All data identified as being part of the firm's data model have a cor-

responding process that captures, validates, maintains, deletes, and uses it.

4. All processing views of the data are accounted for within the data models.
5. References to either data or processes within the individual models are consistent across the models.
6. All data expected by the various processes are accounted for in the data models.

Walk Throughs

Walk throughs are one of the most effective methods for validation. In effect they are presentations of the analytical results to a group of people who were not party to the initial analysis. This group of people should be composed of representatives of all levels of the affected user areas as well as the analysts involved. The function of this group is to determine whether any points have been missed, all the problems have been correctly identified, and whether the proposed solutions are viable. In effect this is a review committee.

Since the analysis documentation should be self-explanatory and nonambiguous, it should be readily understandable by any member of the group. Before the walk through, the group's members should read the documentation and note any questions or areas which need clarification. The walk through itself should take the form of a presentation by the analysts to the group and should be followed up by question and answer periods. Any modifications or corrections required to the documents should be noted. If any areas have been missed, the analyst may have to perform the needed interviews and a second presentation may be needed. The validation process should address the documentation and models developed from the top two levels—the strategic and the managerial—using the detail from the operational level.

Each data element or group of data elements contained in each of these detail transactions should be traced through these models, end to end, that is, from its origination on a source document through all of its transformations into output reports and stored files. Each data element or group of data elements contained in each of these stored files and output reports should be traced back to a single source document.

Each process which handles an original document or transaction should be traced to its end point. That is, the process-to-process flows should be traced and a determination made as to whether all identified inputs are available when the various processes require them. All redundant or ambiguous reports and files should have been noted and eliminated.

Processes should be associated with their owner functions and a note made as to whether any processes are ownerless or are multi-owned. Comparisons should be made between processes to determine whether those with similarities in data needs, time frames, and task content have been combined.

Input/Output Validation

This class of analytical validation begins with system inputs and traces all flows to the final outputs. It also includes transaction analysis, data flow analysis, data source and use analysis, and data event analysis. For these methods each data input to the system is flowed to its final destination. Data transformations and manipulations are examined (using data flow diagrams), and outputs are documented. This type of analysis is usually left to right, in that the inputs are usually portrayed as coming in on the left and going out from the right. Data flow analysis techniques are used to depict the flow through successive levels of process decomposition, arriving ultimately at the unit task level.

Validation of these methods requires that the analyst and user work backward from the outputs to the inputs. To accomplish this, each output or storage item [data items which are stored in ongoing files (also an output)], is traced back through the documented transformations and processes to their ultimate source. Output-to-input validation does not require that all data inputs be used; however, all output or stored data items used must have an ultimate input source and should have a single input.

Input-to-output validation works in the reverse manner. Here, each input item is traced through its processes and transformations to the final output. Validation of output to input, or input to output, analysis looks for data items that have multiple sources as well as those that are acquired but not used.

Data-Source–to–Use Validation

Data-source–to–use validation seeks to determine whether the data gathered by the firm is

1. Needed
2. Verified or validated in the appropriate manner
3. Useful in the form acquired
4. Acquired at the appropriate point by the appropriate functional unit

5. Complete, accurate, and reliable
6. Made available to all functional areas which need it
7. Saved for an appropriate length of time
8. Modified by the appropriate unit in a correct and timely manner
9. Discarded when it is no longer of use to the firm
10. Correctly and appropriately identified when it is used by the firm
11. Appropriately documented as to the type and mode of transformation when it does not appear in its original form
12. Appropriately categorized as to its sensitivity and criticality to the firm

This method of validation is approached at the data element level and disregards the particular documents which carry the data. The rationale here is that data, although initially aggregated to documents, tend to scatter or fragment within the data flows of the firm. Conversely, once within the data flows of the firm, data tend to aggregate in different ways. That is, data are brought together into different collections and from many different sources for various processing purposes.

Some data are used for reference purposes, and some generated as a result of various processing steps and transformations. The result is a web of data which can be mapped irrespective of the processing flows. The data flow models and the data models from the data analysis in the various phases are particularly useful here. The analyst must be sure to cross-reference and cross-validate both the data model from the existing system and the data model from the proposed system.

The new data model should contain new data which must be collected, old data which is transferred intact, and old data which is transformed in some manner from the old to the new model. All data proposed in the new model should be justified, not only from a business need perspective, but also from a cost of acquisition, processing, and storage perspective as well. That is, the analyst must determine whether these costs are justified by the value of the data, both old and new, to the firm.

Consistency Analysis

Consistency analysis for data seeks to determine whether

1. All data elements have been appropriately named and defined
2. All transformations have been identified and described

3. All appearances of the data element have been noted and documented
4. Data elements have been defined in the proper manner and whether the documented transformations are correct
5. All calculations and data derivations have been identified and correctly defined and documented

Consistency analysis for processes seeks to determine whether

1. All processes link together appropriately
2. All processes are being performed in the proper time frame
3. The resources necessary for the process are available when needed
4. All proposed processes are consistent with the job descriptions of those expected to perform them
5. All proposed processes are consistent with the functional descriptions or charters of the organizations to which they have been delegated
6. The people charged with the responsibility for performance of the processes have the necessary authority
7. The supervisory personnel understand the processes in the same way that the persons actually performing them do

Overall, consistency analysis seeks to achieve an end-to-end test of the analysis products, looking for inconsistent references; missing data, processes, documents, or transactions; overlooked activities, functions, inputs, or outputs; etc.

In many respects consistency analysis should be conducted in conjunction with the other validation processes. The analyst and the user alike should be looking for "holes" in the document, not, as with the other techniques, to determine "correctness" but completeness.

Level-to-Level Consistency

Depending upon the size and scope of the project, the analysis may have been conducted in levels. That is, the analysis team may have first tried to achieve a broad-brush overview of the user function. This may have been followed by a more narrowly focused analysis of a particular function or group of functions, and working in successively finer and finer detail, finally arriving at the operational task level. This approach usually works each level to completion before beginning the next lower level. Although in theory the work at the level under analysis should be guided by the work at the preceding (higher) level, in practice, this may not be true.

The reality is that these levels may be under concurrent analysis or may be analyzed by different teams. The latter is especially true in long running projects where staffing turnover, transfer, and promotion continually change the makeup of the analysis team.

Level-to-level consistency validation seeks, in a manner similar to the traditional consistency analysis, to ensure that the information, perspective, and findings at each level correspond, or at least do not conflict with, the information, perspective, and findings at both higher and lower levels.

Carrier-to-Data Consistency

Carrier-to-data consistency focuses on the data transactions and ensures that the data on the documents (data carriers) which enter the firm are passed consistently and accurately to all areas where it is needed.

This phase of the validation seeks to determine whether the firm is receiving the correct data, whether it understands what data it is receiving (i.e., what the transmitter of that data intended), and whether it is using that data in a manner which is consistent with the data's origin.

Since, with few exceptions, the firm collects data on its own forms, and, with even fewer exceptions, the firm can specify what data it needs and in what form it needs that data, this level of validation can compare the firm's use of the incoming data with the data received to ensure that the forms, instructions, and procedures for collection or acquisition, and dissemination are consistent with the data's subsequent usage.

Process-to-Process Consistency

Process-to-process consistency seeks to trace the flows of processes and ensure that the processes are consistent with each other. That is, if one process is expecting data from another, then they both must have a common understanding of the data to be transferred. Process-to-process consistency validation seeks to answer the following questions.

1. Are the process time constraints similar between and across processes, and are all processes aware of those constraints?
2. Have all processes been identified?

3. Are there accompanying narratives for each process as a whole and for each component task?
4. Are all inputs, outputs, data storage and retrieval activities, and forms, reports, and transactions for each process clearly identified and described?
5. Have all forward and backward references for these items been clearly indicated? Do those references have corresponding references in the sourcing or receiving process?

Other Analytical Techniques

Zero-based analysis

Zero-based analysis is similar in nature to zero-based budgeting, in that it assumes nothing is known about the existing environment. It is a "start from scratch" approach. All user functions, processes, and tasks are reexamined and rejustified. The reasons for each user activity are documented, and all work flows are retraced.

Zero-based analysis is especially necessary for projects in areas undergoing resystemization or reautomation. Here the existing systems and automation may have been erroneous, or the environment may have changed sufficiently to warrant this start-from-scratch approach.

The analyst should never assume that the original reasons for collecting or processing the data are still valid. Each data transaction and each process must be examined as if it were being proposed for the first time or for a new system. The analyst must ask

1. Does this need to be done?
2. Are all the steps correct? And, are they all necessary?
3. Should this task or process be performed by the current unit?
4. Does the work accomplished justify the resources being devoted to it?
5. For each report
   - Is this report necessary?
   - If it is necessary, is it necessary in its current form?
   - Is it still necessary at its current level of detail?
   - Should it be produced as frequently as it is?
6. For each process
   - Is it still necessary?
   - Does it need to be performed as frequently as it is?
   - If the process is manual should it be automated, and if currently automated, should it revert to manual?
   - Do the procedures and standards which govern the process still ap-

ply, or should they be revised? Made simpler? Made more comprehensive?
- Can it be combined with other similar processes?
- Has it grown so complex that it needs to be fragmented into a larger number of more simplified processes?
- Can the cost of the process be justified in terms of the benefit to the firm?
- Does the volume of work expected justify the size of the organization or the resources devoted to it?

Data event analysis

This modification of transaction analysis assumes that all business activities are triggered by data events or data transactions. These data events or data transactions are stimuli from either internal or external sources which constitute the day-to-day business of the firm.

These stimuli come from a variety of sources and cause the business to react in predictable ways. These data events may be manual or automated, or in some cases may result from the passage of time or from some other internal activity. There may or may not be any physical notification of the event. In the absence of any physical data carrier, or even if one is present, the analyst must determine

- What the firm needs to know about the event
- How that knowledge may be validated
- What the firm should do with the information
- Where and how much of it must be stored for future use

These actions and determinations should be analyzed irrespective of the persons or areas which may actually respond and of the format and content of the actual event notification. Each data event results in a data activity flow. A data activity flow consists of a series of data handling activities which are concerned purely with the receipt, validation, and movement of the data, and not with the processing or manipulation of that data. Data event flows may incorporate multiple processing steps or multiple processes.

Each data event flow starts with a data trigger of the firm from its initial identification or recognition and is traced from its source through its eventual storage in the files of the firm. Data activities are limited to

- Receipt, identification, or recognition
- Retrieval of any previously stored operational or reference data

- Verification of the contents of the data trigger
- Adding, updating, or deleting the items of interest from the data trigger to the previously acquired data within the firm
- Archive any previous occurrences of the same data
- Store the new data in the firm's files

Methods and procedures analysis

Although the tendency is for data processing professionals to assume that all solutions to user problems must be automated data processing solutions, this is not always the case. The user's problems are business problems, not necessarily data manipulation or presentation problems. This methodology includes generally accepted principles of business systems analysis and integrates both automated and manual analysis, and automated and manual solutions. It analyzes the user's functions, processes, activities, and tasks and includes manual-to-manual, automated-to-manual, manual-to-automated, and automated-to-automated flows.

General to the specific analysis

The top-down approach to analysis and design requires that its first phases be, of necessity, rather general. That is the functions, processes, and data are treated in the abstract, as generic forms and as aggregates, rather than as specific tasks. In other words, rather than attempting to dive right into the detail of the tasks, the analyst should first try to put together less detailed (although not necessarily less complex) pictures of the firm, and as knowledge and understanding increase, to create successively more detailed pictures.

The analyst first attempts to gain an understanding, or overview, of the entire firm, the enterprise view, and from there proceeds to more specific and thus more detailed, views of individual user areas and activities. These detailed views are always developed within the enterprise framework. The enterprise framework is used to guide the development of and to validate the specific, detail views.

Part

4

Analysis Issues

The objective of all dedicated employees should be to thoroughly analyze all situations, anticipate all problems prior to their occurrence, have answers for these problems, and move swiftly to solve these problems when called upon. . . .

However, when you are up to your ears in alligators, it is difficult to remind yourself that your initial objective was to drain the swamp.

Chapter

15

Manual and Automated Systems Examination

CHAPTER SYNOPSIS

Once the determination has been made to develop an automated system, the analyst must evaluate the various issues of automation. These include among others: Should micro, mini, or mainframe systems be developed? Should they be online or batch? Should they be stand-alone or integrated? Should the firm build the application in-house or should it attempt to buy a package commercially? Other issues include resident machine size, package evaluation, and selection.

This chapter discusses the various parameter decisions, tradeoffs, advantages, and disadvantages for the issues discussed above. In addition, where appropriate, there are lists of questions and issues which need to be addressed.

Top-Down versus Bottom-Up Analysis

Top-down analysis

Top-down analysis is a term used to describe analysis which starts with a high level overview of the firm and its functional areas. This overview should be a complete picture of the firm, but it should be general rather than very detailed. Once the overview has been developed, the analyst produces successively more detailed views of specific areas of interest. This process of developing more and more detailed views is called decomposition.

Top-down analysis is the method used by most major commercial methodologies and is considered to be the most thorough form of analysis. The order of the life cycle phases discussed earlier is based on a top-down analysis.

The difference between top-down and bottom-up analysis can be illustrated in the following manner (see Figure 15.1). Top-down analysis takes a finished product and attempts to find out how it works. The product is taken apart; the atomic parts used to create it are examined and documented, subassembly by subassembly. Bottom-up analysis starts with the gathering of all the atomic parts; the analyst then attempts to figure out what they will look like when they are all assembled and what the completed product can and should do.

The top-down method of analysis is usually accomplished in a phased manner. Many times the various phases are worked on by different teams, or by different people from the same teams. The work is assigned to correspond to the perceived skill levels required (Figure 15.2). The senior analysts work on the higher levels, while the junior analysts typically work on the lower levels. The advantages of top-down analysis are

1. The detailed analysis tends to be more complete and provides greater opportunity for identification of the interweaving of processing threads.
2. Duplication of activity, overlapping function and processing, and

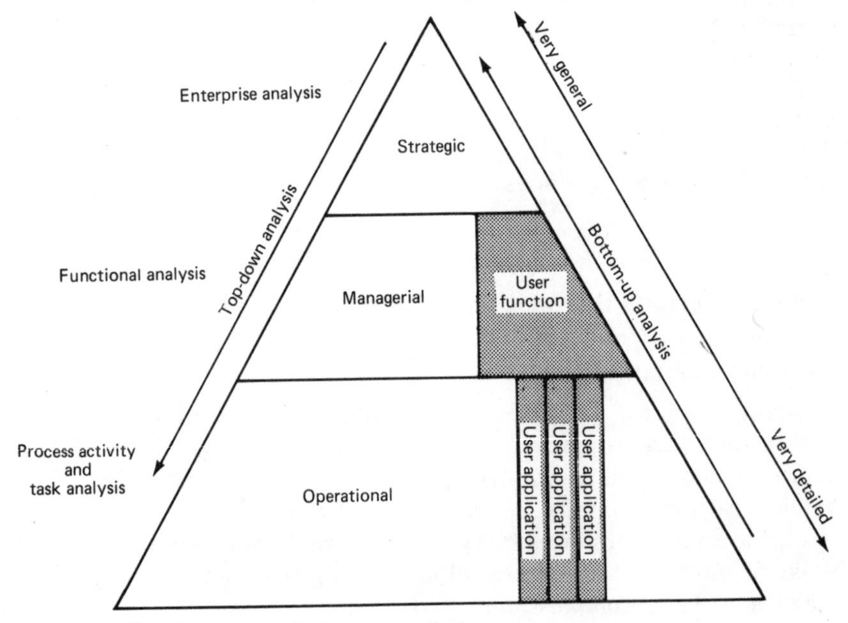

Figure 15.1 Top-down versus bottom-up analysis.

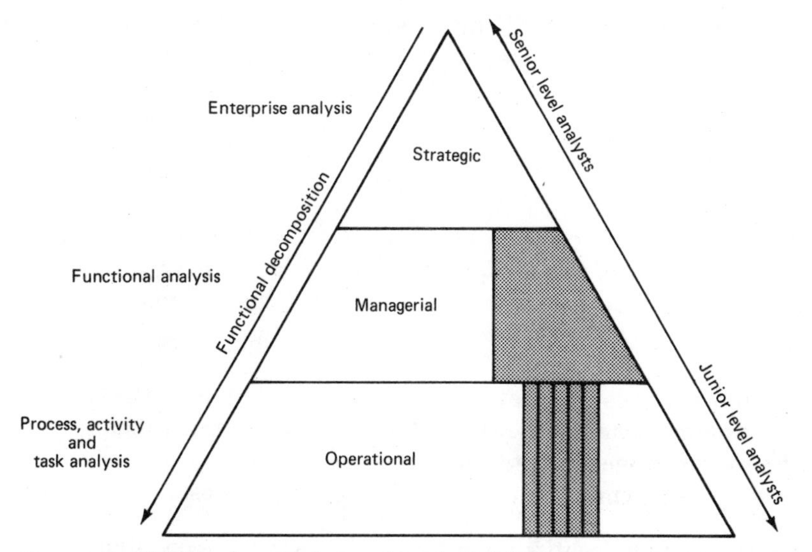

Figure 15.2 Skill levels needed at each analytical level.

inconsistency of activity are more readily apparent when looking at overviews than they are when starting with the detail level.

3. Top-down tends to provide more perspective and to highlight problems of organization and overall work flow, as well as opportunities for work flow streamlining.
4. Top-down tends to highlight overall data usage and data needs more easily than does the bottom-up approach.
5. Once completed the overview analysis can serve as the basis for many differing application development projects, and it usually requires little more than periodic updating.

The disadvantages of top-down are

1. It tends to be more difficult and time consuming, and thus more expensive. This is due to the additional levels of analysis and the additional work.
2. It looks at areas outside that of the user-sponsor, which is difficult for the user-sponsor to deal with *and fund.*
3. The benefits to the user-sponsor tend to be less obvious and longer in coming than with bottom-up.
4. Top-down tends to require more contact, support, and information from senior management, since the highest levels of analysis concentrate at their level. Senior managers sometimes express impa-

tience with this kind of "fishing" activity, regarding it as something with little relevance which could be dealt with at lower levels.

Bottom-up analysis

Since many application projects are very specific in their focus and are operational in nature, the analysis for these projects may start with the clerical or operational activities which are their primary focus. Here the functions are well known as are the problems and user requirements.

Bottom-up analysis and development is aptly suited to the operational environment and is the favored method for organizations in the first and second stages of data processing growth.

Bottom-up development has the following advantages.

1. Since the work is localized and focused, it is much more limited in the early stages. Bottom-up projects are able to home in quickly on satisfying the user-sponsor's needs.
2. Being more focused, bottom-up development tends to stay within the bounds of the user area, further limiting the amount of work necessary. This limitation on work makes bottom-up projects faster and less expensive.
3. It is more closely suited to user-initiated user-specific application development projects.

The disadvantages of bottom-up projects are

1. Their limitation of scope tends to preclude activities which cross user process and functional boundaries. This also limits the analyst's ability to identify and correct processing redundancies and data usage anomalies.
2. Because the focus of any given project is narrow, there tends to be significant rework in related projects as the incomplete pictures of previous analyses are updated and viewed from the differing perspectives of new user areas.
3. Since bottom-up analysis focuses on the operational areas of the firm, the analyst's view is both narrow and vertical; it does not permit the analyst to view the impact of the particular operational area on other operational areas.

The advantages and disadvantages of top-down and bottom-up analysis are presented in Figure 15.3.

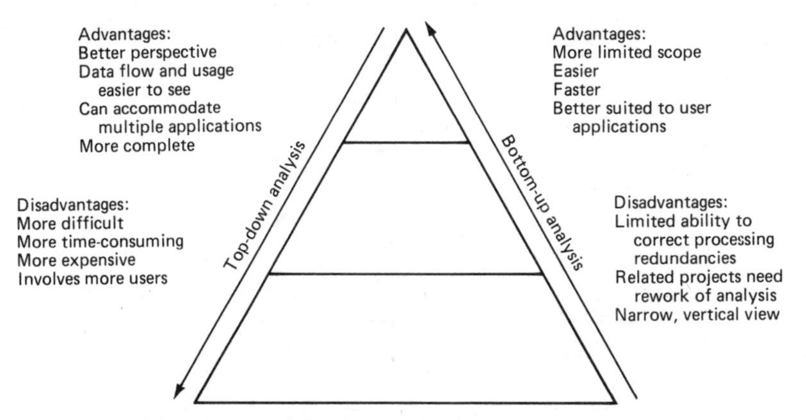

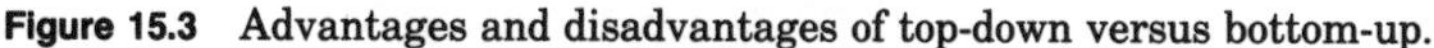

Figure 15.3 Advantages and disadvantages of top-down versus bottom-up.

Online versus Batch Systems

Early systems development, being limited by both technological availability and analytical experience, tended to duplicate the sequential batch processing which itself was a holdover from the early manufacturing experiences. Work flows were treated as step-by-step processing governed by strict rules of precedence.

The methods of data entry and automated input required strict controls to ensure that all inputs were received and entered properly. These controls were necessary because of the time delays between acquisition of the source data and their actual entry into the automated files.

Additionally because of the number of processing steps, such as keypunching, sorting, and collating of the data, occurring prior to the actual machine processing, the possibility and probability of data being incorrect or of items being lost were rather high. To alleviate these problems, analysts developed additional manual and automated steps and inserted them into the processing streams for collecting input items into groups, called batches, developing control totals on both the number of items and the quantity and dollar amounts of those items. As each batch was collected and verified it was processed against the master files. This type of processing tended to become somewhat start-stop in nature. Batches tended to be processed together, which required further controls on the number of batches and on overall batch totals. See Figure 15.4 for a description of the sequence of batch processing.

Since processing could not be complete until all batches were processed, any activities on the file as a whole, or on all transactions, tended to wait for the last batch. In some cases, the batch processing only verified the inputs, and transaction-to-file processing waited until all batch work was completed. This was necessary because the master files were usually maintained in an order different from that of the

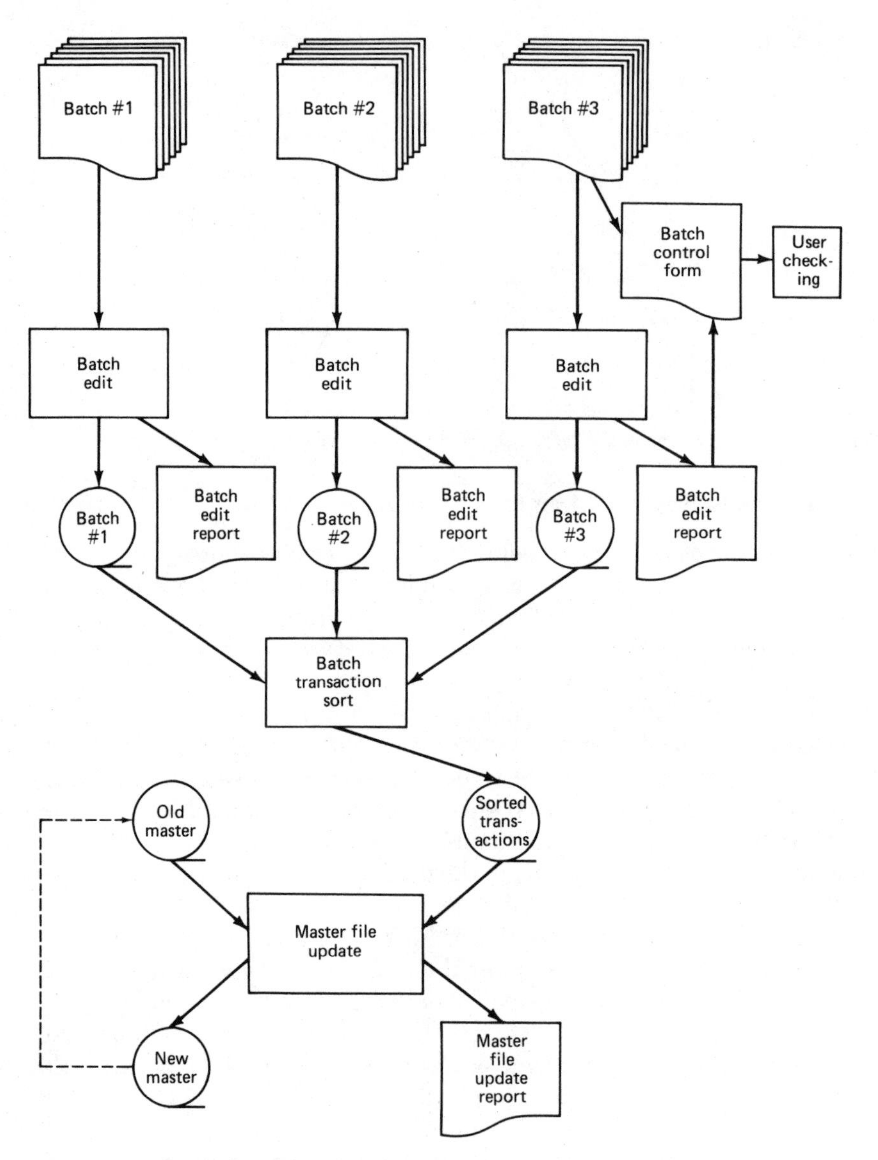

Figure 15.4 Batch processing sequence.

randomly collected transactions. These randomly accumulated transaction files themselves had to be sorted into the same order as the master files before processing.

This mode of processing became so ingrained into the development mentality that initial online processing continued to mimic batch processing, in that groups of transactions were aggregated and entered on a screen by screen basis.

Batch transaction processing is suited to both sequential and direct processing; however any environment where the files are maintained in a sequential medium mandates that all processing be in batch form. Where the master files are maintained in random-access–based files, true online processing can occur.

Online processing is usually characterized by transaction-at-a-time designs, where the transaction data directly updates the master file in a random manner (Figure 15.5). In this mode, the user is presented with a screen which allows the entry of a single transaction of data. That data is verified independently and applied to the master file. Online processing is random in nature and is based upon transaction arrival, while batch processing waits for a sufficient number of transactions to arrive to make up a batch.

Batch processing takes advantage of the fact that task setup usually takes as much as if not more time than does the actual processing. Thus if multiple transactions can be processed with one setup, time will be saved. This is the assembly line theory. On the other hand, online is

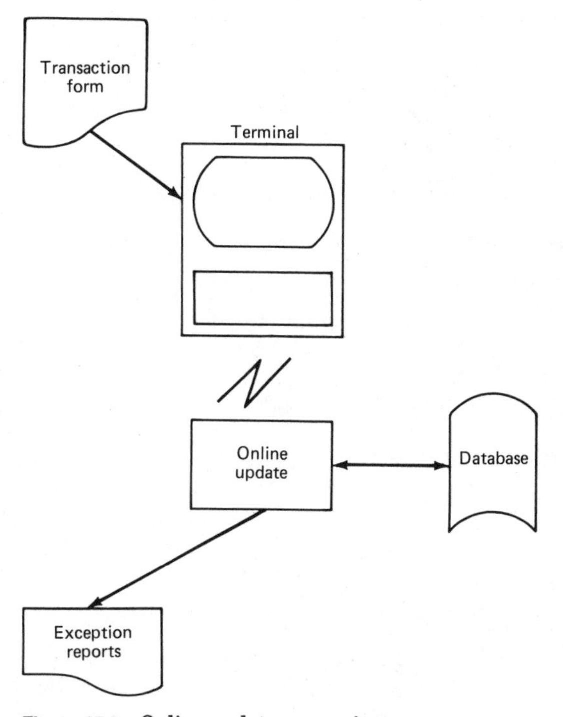

Figure 15.5 Online update processing.

similar to the artisan method where one person performs a complete sequence of tasks.

In many cases however, batch processing is required by business rules and policies. For instance

1. The business may dictate that certain orders have priority when going against inventory. Since orders are received on a random basis during the day, it is necessary to collect all orders, and just before going against inventory, sort them into priority order.
2. In demand deposit accounting, the business rules usually state that all deposits are processed first, followed by any special instructions (i.e., stop orders) followed by certified checks, followed by normal checking activity. Again since these transactions are received randomly during the day, they must be accumulated, and sorted and processed in the correct sequence at the end of the day.

Issues Affecting the Online versus Batch Decision

1. Any business rules or conditions which necessitate either batch or online processing
2. Any transaction type priorities which may affect the decision
3. Any user needs for rapid access to data during the working day
4. Any data movement problems
5. User proximity to the processing center or to other users
6. User computer sophistication or computer literacy
7. Any "windowing" requirements or other timing requirements
8. Availability and quality of communications facilities
9. Volume of data to be processed
10. Complexity of the data to be processed
11. Cleanliness of the data to be processed
12. Any resource constraints which preclude either online or batch processing

Mainframe, Mini, and Micro Systems

Prior to the late 1970s and early 1980s, the choice of automated system implementation environment was limited to centralized hardware, which bore the labels "mainframe" and "minicomputer." These labels usually referred to distinctions in both size and power. The "mini" label was usually applied to that hardware which was obtained for stand-alone or "turnkey" systems. A turnkey system was a stand-alone system which was acquired as a complete package of hardware and software.

As the performance of systems increased in terms of both throughput and capacity, the labels referred to the size of both of the machines themselves and to the vendors who manufactured them. Generally, minicomputers were manufactured by the smaller hardware vendors. What these boxes had in common, however, was their need for special conditioned rooms and operations staffs. As the technology of the manufacturing process improved, even the size distinction blurred. "Mainframe" became the label for very large powerful boxes and "mini" became the label for all others. Figure 15.6 shows mainframe price, size, and performance curves.

With the advent of the micro- or desktop computers, new, previously infeasible applications and uses for computers became practical. "Micro" is the label applied to that machine known variously as the personal workstation, personal computer, or desktop. These machines, which originally were little more than glorified calculators or intelligent terminals compared to the mini and mainframe, and which were

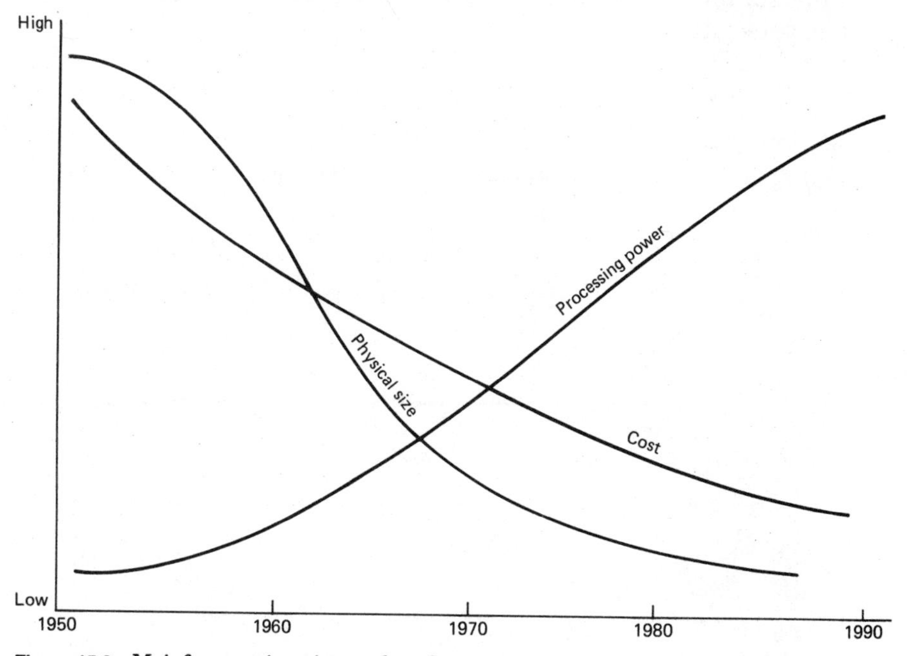

Figure 15.6 Mainframe price, size, and performance curves.

isolated from the mainstream data processing environment, were designed to run packaged products, such as word processing, database, and spreadsheet packages.

The versatility and relatively low cost of these small machines have made them ideal for user applications. While small in size, they have achieved extensive power and capacity. The evolution of the mainframe and minicomputers took about 25 to 30 years to reach their present state. Figure 15.7 shows microcomputer price, size, and performance curves. The microcomputers by contrast have taken less than 10 years to reach a point where they rival their larger cousins in terms of speed, capacity, and availability of software. In the next 10 years, one can expect that these microcomputers will exceed all but the largest and most powerful supercomputers in speed and capacity and will have data storage capability to rival many present-day machines. Figure 15.8 is a comparison of microcomputer versus mainframe price-performance curves.

The rapid development of these very small machines has opened up new areas of automation within companies and has placed many firms back into the first stages of a new data processing growth cycle. Additionally, applications which were previously only available on mainframes have been made available on the micros, leading to intense reautomation efforts in an effort to take advantage of these inexpensive, personal machines.

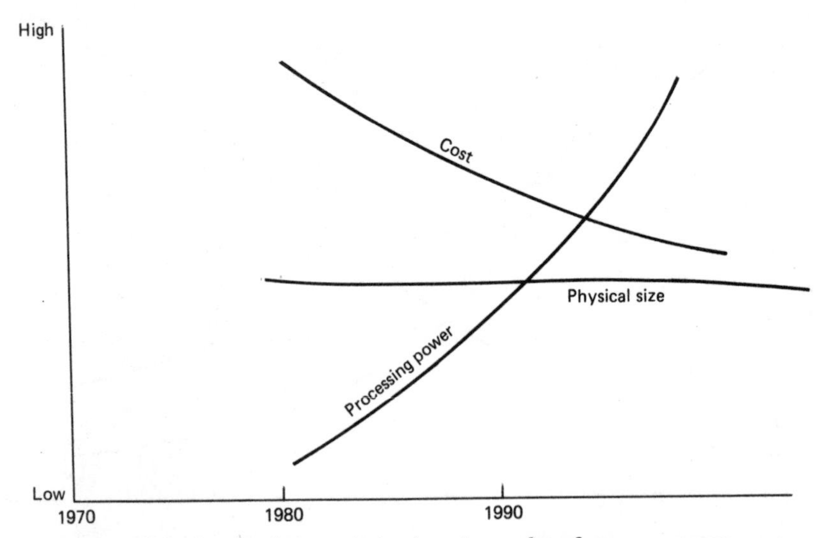

Figure 15.7 Micro (personal computer) price, size, and performance curves.

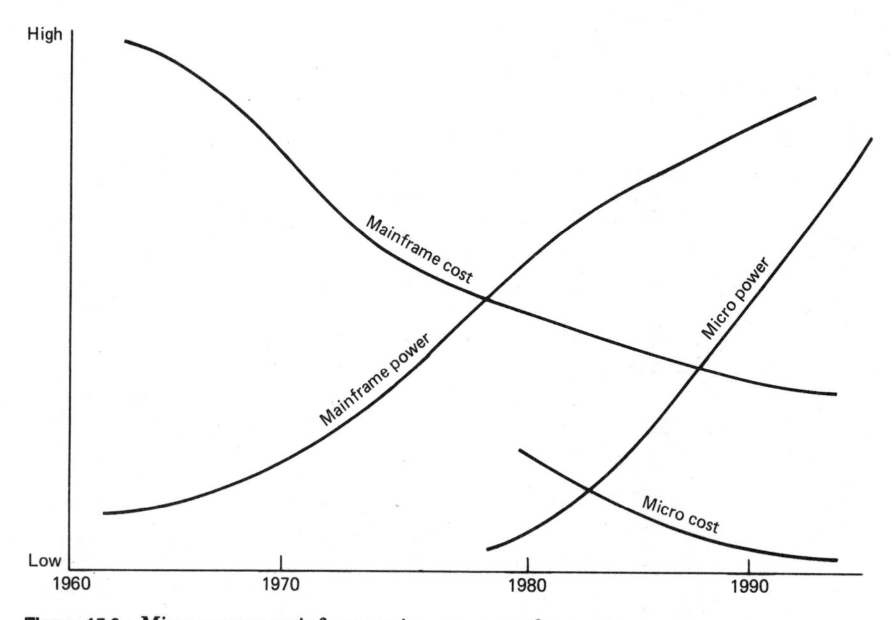

Figure 15.8 Micro versus mainframe price versus performance curves.

The development of packaged applications for record keeping and analysis and for routine and highly specialized business processing support has made these machines relatively common office tools. As their capacity and speed increase and as their cost decreases, more and more applications will be found for them. The analyst must seriously consider this new and wide range of machinery when looking to create a practical business solution for the client-user area.

Although microcomputers were originally designed as stand-alone, personal machines, the tediousness of manual data entry has caused both the business and personal user communities to demand and get the capability to move data to and from the mainframes on a direct, automated basis. Although, currently there are format and speed restrictions, it is conceivable that in the very near future data will move freely and quickly between the two environments, opening up vast opportunities for cost-effective automation for the user areas. The dual mode capability, local and remote, plus the growing ability to network, that is, interconnect, these machines, many of them with common libraries and common data storage, will further open up these machines to application use. Figure 15.9 shows the multiple modes of microcomputer environments.

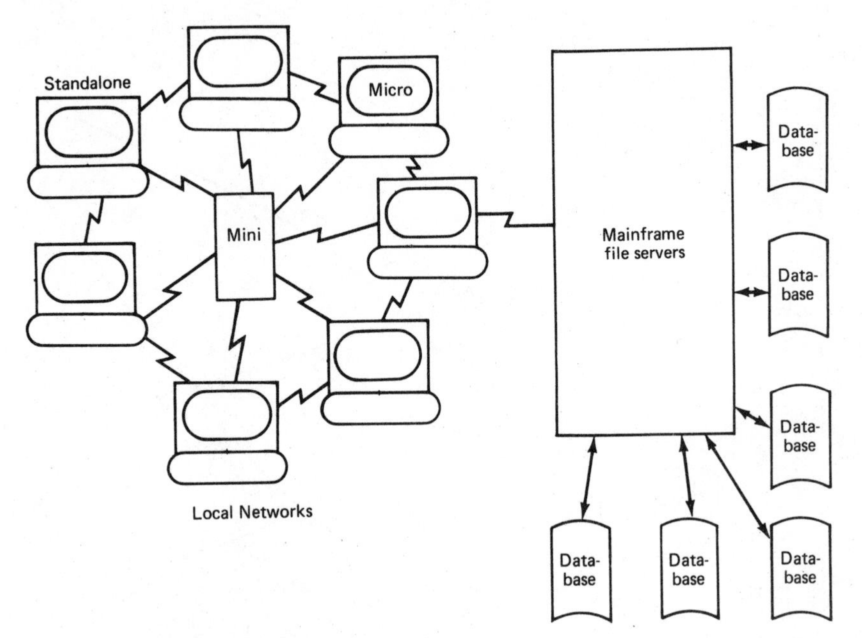

Figure 15.9 Multiple modes of micro environments.

Micro, Mini, or Mainframe Issues

1. Volume of data
2. Size of ongoing files
3. Type of processing
4. Number of potential users
5. User location
6. Estimated length of processing cycle
7. Existing mainframe capacity
8. User sophistication and computer literacy
9. Special software or hardware requirements
10. Reporting volumes
11. Type and location of existing hardware
12. Internal expertise
13. Any data sharing requirements
14. Data entry volumes
15. Any special communications requirements
16. Processing complexity

Integrated Systems versus Stand-Alone

Integrated systems are those which attempt to look at the corporate environment from a top-down viewpoint or from a cross-functional and cross–business-unit perspective. To illustrate, an integrated system would be one which looks at human resources, rather than treating payroll and personnel as separate processes, or at general ledger rather than at balance sheet, accounts payable, and receivables, etc. Integrated systems are modeled along functional, business, and strategic lines rather than along process and operational lines. Figure 15.10 compares stand-alone and integrated systems.

Integrated systems recognize the interdependency of user areas and try to address as many of these interrelated interdependent areas as is feasible. Integrated systems are usually oriented along common functional and data requirement lines. Integrated systems require top-down analysis and development since it is easier to determine overall requirements, and also because integrated systems development makes it necessary to understand the interdependencies and interrelated nature of the various applications which must be hooked together to achieve integration. The scope and requirements of integrated systems are difficult to analyze and generally require more time to develop.

Since integrated systems cross functional, and thus user, boundaries, many user areas must be involved in both the analysis and subsequent design and implementation phases. A multiuser environment is much more difficult to work with because even though the system is integrated, the users normally are not. Each user brings his or her own perspective to the environment, problems, and requirements, and these differing perspectives may often conflict with each other. The analyst must resolve these conflicts during the analysis process, or during the later review and approval cycles. In addition to the conflicting perspectives, there are normally conflicting system goals, and, more important, conflicting time frames as well.

Stand-alone systems, by contrast, are usually those which are self-contained and are designed to accomplish a specific process or support a specific function. Stand-alone systems are usually characterized by a single homogeneous user community, limited system goals, and a single time frame.

There are no guidelines which distinguish stand-alone systems from integrated ones. In fact, stand-alone systems may also be integrated in nature. There are also no size or complexity distinguishing characteristics.

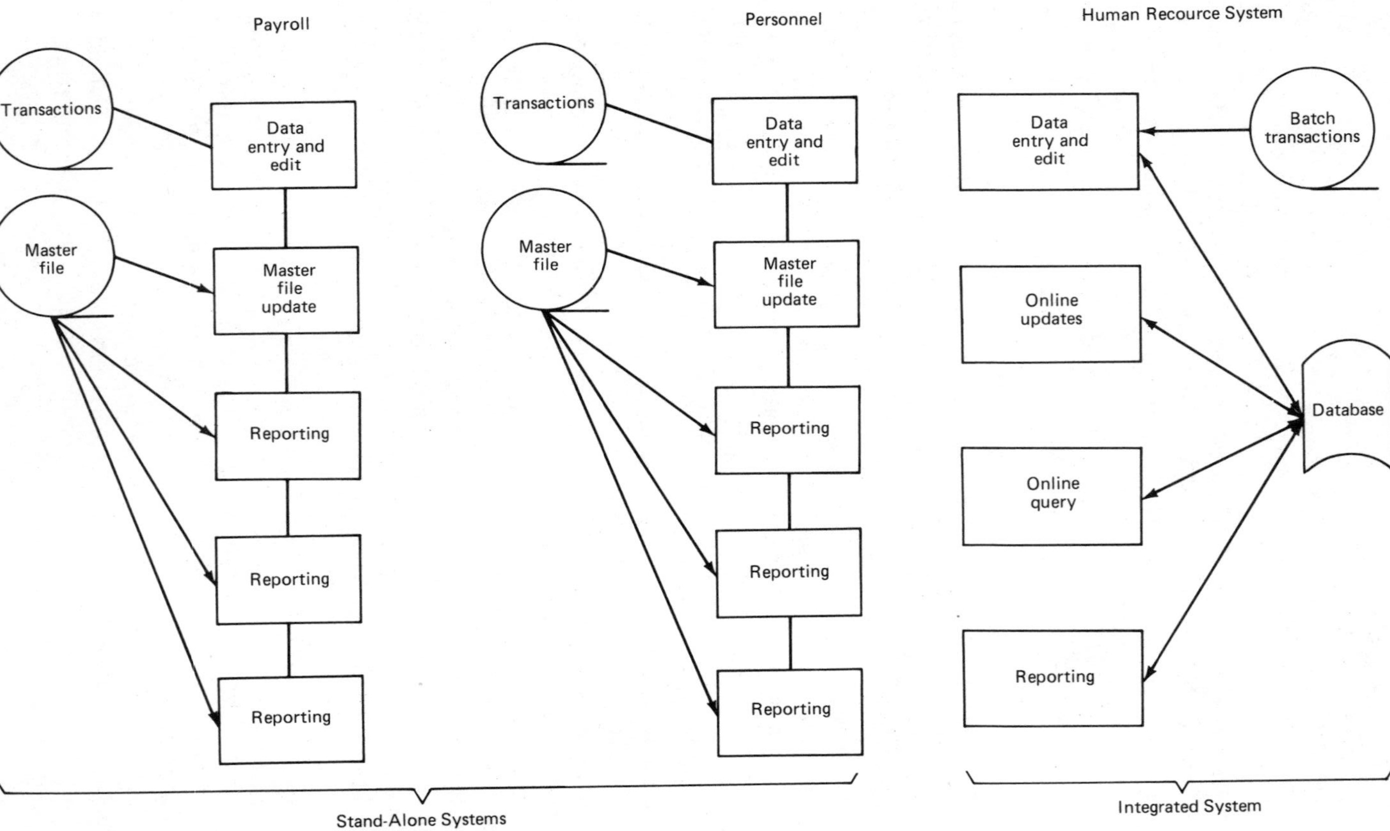

Figure 15.10 Stand-alone versus integrated systems.

The decision as to whether to attempt to develop an integrated system or to develop a stand-alone system is dependent upon the following issues.

1. The location of the firm on the growth cycle. Firms in the early stages of the cycle should not attempt integrated systems, whereas firms in the late stages should always strive for them.
2. The ability to assemble all interested and relevant users, and the ability to get them to agree to participate, to compromise where necessary, and to jointly fund the project.
3. The commitment from user management to devote the time and resources to a complete top-down analysis.
4. The ability to get all users to share the stored data and the tasks of data acquisition and maintenance.
5. User understanding of the processes and functions which the system would ultimately be designed to service.

Make-versus-Buy Decisions

Once the process of analysis has been completed and a proposed systems solution (*not* the design itself) for the user environment has been developed, one last set of tasks faces the analyst, that is, determining whether it is feasible to obtain a prepackaged application system from a vendor as opposed to developing a completely new system using internal resources. This analysis is called the "make-versus-buy" decision. In many cases obtaining a prebuilt package is extremely cost-effective. The make-or-buy decision may also extend to retaining an outside firm to custom-build a package to the user's specifications.

Make-versus-Buy Issues

1. Costs associated with developing the system in-house
2. Costs associated with acquiring a system externally
3. Any exceptional or specialized requirements which are unique to the firm
4. Number of externally available packages
5. Time frame in which the user needs the system

6. Development time estimates versus time to evaluate, select, install, and modify an external package
7. Availability of internal personnel to develop the needed system in-house
8. The degree of comfort the analyst and user have in the firmness of the specifications for the system
9. The expected life of the proposed system
10. The presence of any proprietary information about the firm's operations or the user's system, which the firm may be unable or unwilling to release to outside firms or nonemployees
11. Expected volatility of proposed system
12. Absence of specialized development or functional expertise within the firm which might be needed to developed the proposed system

Package Evaluation and Selection

If the decision has been reached to buy, the analyst must recognize that any system that is acquired, rather than custom-built, will not fully meet the firm's needs. Prebuilt packages are by their very nature generalized to suit the largest number of potential customers. This implies that while most packages will address the basic functional requirements, some percentage of the needed user functionality will be missing. Each package will have its own configuration of supported functions and these may not be the same from package to package. Those functions which are supported, basic or otherwise, can also be implemented in sometimes radically different ways, and the depth and comprehensiveness of the functional support can also differ radically.

Additionally there may be some specialized company functional needs which may not be addressed by any vendor package. The analyst can also expect that each package will support not only different functional requirements but may also be designed to operate in markedly different types of businesses. For instance a financial package designed for manufacturing organizations will have different design characteristics from one designed for a financial or service organization. This difference in functionality may make comparative product evaluation difficult if not impossible.

Most packages were originally designed for a specific company in a specific industry and were then generalized for commercial sale. These packages may have been designed by the specific firm itself or designed for the firm by an outside service consultant. Depending upon its origins, the implementation may vary from very good to very poor, and the documentation can be expected to vary a great deal. In any case, because of their origins as custom systems, one can expect the implementation to bear a strong imprint of the original users.

Industry surveys indicate that a functional and procedural "fit" of between 30 to 40 percent is considered average. This means that 30 to 40 percent of the package capability will exactly match the company's requirements. The analyst must assess the closeness of the fit between desired and needed functionality. The analyst must also assess the closeness of the procedural implementation to the way the firm currently does business and assess the impact of either modifying the firm to conform to the package requirements or modifying the package to conform to the firm's requirements.

In addition, the analyst must "look beneath the covers" at how the system works, not only what it does. Many packages come with their own forms, coding structures, processing algorithms, and built-in standards and policies. Many of these package internals are not changeable, and the analyst must determine the degree to which the user is willing to accept them (Figure 15.11). The analysis must also examine in detail the vendor of the product, the vendor's service and reliability, and any vendor restrictions on the company's use of the product. The lack of available literature and of available detailed evaluation information may make this evaluation process very time-consuming.

Examining the Buy Option

Some issues to consider when examining the buy option are

1. The system's maintainability
2. The availability of training

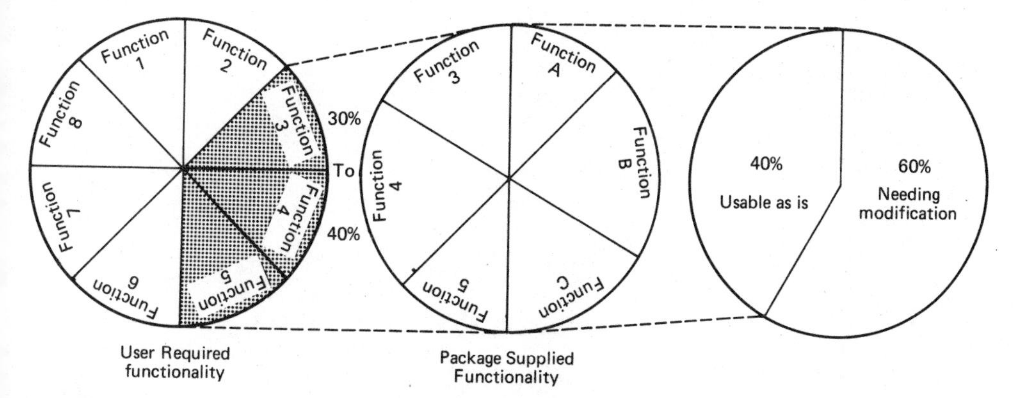

Figure 15.11 Package application usability analysis.

3. The level and quality of the documentation
4. Viability of the vendor
5. The experiences of other customers
6. Frequency of vendor maintenance and modification
7. Vendor's support capability, including problem resolution, "hot-line" service, etc.
8. The ease with which the system can be enhanced by internal personnel
9. The impact of any company modification on any system warranties, guarantees, or service contracts
10. The relative currency of the software and hardware base of the package
11. Any related offerings by the same vendor
12. Any items or functions promised for future delivery
13. Testing and benchmark periods
14. Acquisition options, i.e., lease, franchise, license, purchase
15. The impact of any proprietary products or processing on the company
16. Any restrictions on disclosure, resale, etc.
17. Site licensing versus single Central Processing Unit (CPU) licensing versus companywide licensing
18. Duplication restrictions on software, documentation, or manuals
19. The willingness of the vendor to "customize" the package to the firm's specifications

Chapter

16

Database Environment Considerations

CHAPTER SYNOPSIS

This chapter provides a general discussion of databases, data structure forms, DBMS technology, and their impact on the analytical tasks. A discussion of the roles of the database administrator and the data administrator in the analysis of a database-based system are also included.

What Is a Database?

A *database* is a collection of data needed to support and record the business of the firm. These business records include the ongoing records of the firm, the day-to-day business transactions, and any material or information which is used for reference purposes. A database has the following characteristics.

- It is a base of data.
- It is a common pool of data.
- It could be manual or automated.
- It is an orientation or a frame of mind.
- It is a frame of reference for systems development.
- It can be viewed as a set of "file cabinets"

where

Each cabinet contains a number of "drawers"

Each drawer contains a number of indexed "folders"

Each folder contains a number of related "records"

Each record contains a number of related items of information or data

What Is a Database Management System (DBMS)?

Most database management systems have been developed either by the hardware manufacturers themselves or by firms which are exclusively devoted to the development and marketing of software. In some rare instances, database management systems have been developed by private firms for internal use and have then been generalized and marketed in much the same way as packaged application systems.

Generally speaking a database management system consists of

- An extensive collection of modules, programs, and tables
- An access method and an access methodology
- A set of data manipulation and retrieval tools
- Built-in provisions for data integrity and security
- A set of file, record, and element descriptions
- Rules of logic for file construction and data handling
- Specifications for physical data storage

Why Use a DBMS?

The uses of a DBMS are

- Facilitates the removal of external data manipulation routines from program streams (i.e., sorts, merges, etc.)
- Facilitates the elimination of master-in/master-out logic which characterizes most sequential-processing batch-oriented systems
- Eliminates the need for full file processing in selective retrieval or update conditions
- Facilitates concurrent use of the same file by multiple programs (update and/or retrieval)
- Provides for data recovery after failure
- Provides for data access logically rather than physically

Two Types of Databases

Reference

Reference databases are, as the name implies, mainly used for reference purposes. They are rather static in nature in that the data contained within them rarely change.

This type of database is usually designed to contain information specific to a single major data entity of interest to the firm. For example, a database may contain information about customers, or accounts, or orders, or employees, or locations.

These databases are usually designed to be used in common by many diverse applications. They act as the single repository of data about their particular entity. They are sometimes also known as subject-specific databases.

Process related

Process-related databases are similar in nature to the files which are used by the more traditional flat-file sequential-process applications. They are designed to contain data which are usually of a transitory nature and are dynamic in use. That is, their contents are continually changing. Process databases are holding bins for work in process. Data contained there may be held between cycles of an application or pending further input. Data in this type of database are usually incomplete in some way or other.

A Comparison of DBMS Data Structure Options

Database management system products are tools for structuring and managing the application data of the firm. Some firms are in the process of selecting a DBMS for the first time; others are in the process of reevaluating their existing DBMS with an eye toward replacement. Still others have multiple DBMS products in their software portfolio.

The variety of products on the market today offers developers the ability to produce applications using any of three major data structure models: hierarchic, network, or relational. Some products offer more than one of these structural model options.

However, each DBMS has a primary underlying structural model, and thus is more effective with data whose structural characteristics most closely correspond with that specific model. Matching the right DBMS with the structural characteristics of the firm's data can greatly affect the efficiency of the application.

The analyst should develop a data-structure–independent model of the applications data (such as an entity-relationship-attribute data diagram) and translate that model into each of the three structures before evaluating which model is the closest fit.

Aside from evaluating an application's data under each of the three data structure models, the analyst should also have an understanding of (a) the data orientation of each of the different types of business systems and (b) the strengths and weaknesses of the various structural models upon which the DBMS products are based.

Aside from the obvious functional differences between applications, most organizations have many different types of systems, each having different scopes and different levels of interaction between the systems. Generally speaking these systems can be categorized into transactional, analytical, and administrative (Figure 16.1).

Transactional systems

Transactional systems are those which involve the primary record keeping of the firm. They are the replacement systems for the paper work which drives the business. They may be order processing, cus-

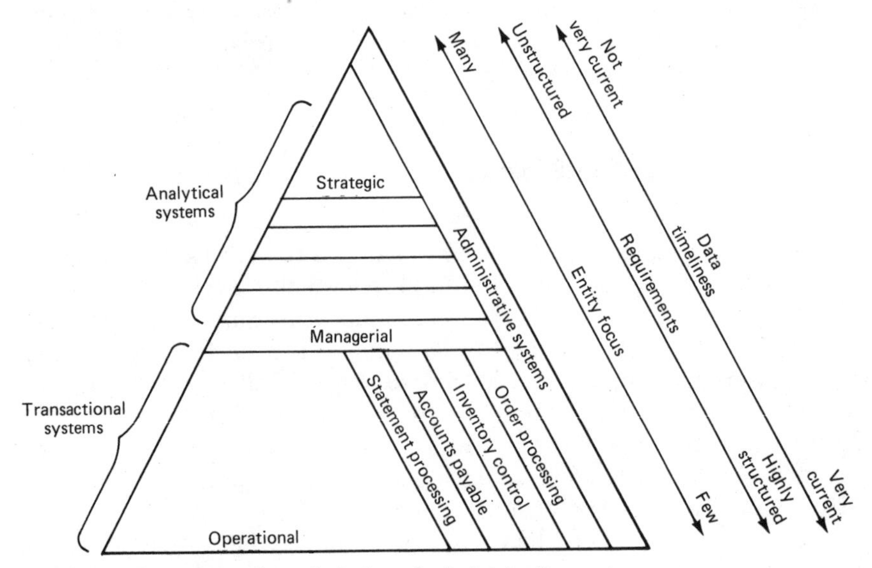

Figure 16.1 Transactional, analytical, and administrative systems.

tomer service, inventory control, statement processing, accounts payable or receivable, or sales support systems. They have in common (a) the need to share data about the major business entities of the firm, (b) well-defined data requirements and procedures, and (c) highly repetitive processes. These systems are the primary data-gathering and data-generating systems of the firm.

Transactional systems tend to center around processing information about one or two entity types at a time (i.e., customers and accounts, customers and customer orders, vendors and products, etc.), and are normally key-driven. Since they service the operational levels of the firm, transactional systems are vertical in nature. That is, they support the processing and provide for the data needs of a specific user area. The operational area usually has an immediate need for data and to be effective must work with the most current information.

Because of their strong focus on individual entities, their fixed processing requirements, and their limited reliance on multiple entity-to-entity relationships, transactional systems tend to work best within the hierarchic and network models.

Analytical systems

Analytical systems also support the business of the firm but are not directly involved in transactions or record keeping. They are usually post facto, in that they are primarily reporting systems that use data which already exist. These data may have been internally or externally generated.

Analytical systems include sales analysis, most financial reports, marketing analysis, and the myriad of systems which have been called management information systems, decision support systems, etc. Into this category we could also place those functions which have come to be called information center or end-user computing systems.

These types of systems tend to deal with sets of entities or with selected data from many different entity types. Since they are not usually transaction-driven, they tend to rely less on occurrence keys and more on the relationships between the various entities. Analytical systems support the managerial and strategic levels of the firm and are horizontal in nature, i.e., they cut across operational functions. As such their need for data is more extensive, more variable, and usually less immediate than that of the operational level, and their need for data currency is usually much less.

Because of their need for data about multiple entities and their need to relate these entities in multiple ways, analytical systems will work best in the network or relational modes; however, they will perform adequately within the hierarchic model as well.

Administrative systems

The final category comprises the administrative systems of the firm. These systems service the firm as a whole and have little to do with the specific business of the firm. Rather they deal with the firm as an entity.

Administrative systems include human resources, payroll, general ledger, and fixed asset control systems. They are more or less standardized from firm to firm; for the most part, they are self-contained systems. They are only tenuously related to other systems, have their own data sources and files, and only rarely require ongoing data from the operational files of the firm, although they may accept feeds from them.

Administrative systems tend to resemble transactional systems in that they gather data and analytical systems in that they have heavy reporting requirements. They differ in that they tend to focus on one entity type at a time (i.e., employees, accounts, offices, warehouses, etc.) rather than on groups of entity types.

Administrative systems are both vertically and horizontally oriented. They are normally firmwide systems and provide data and support to all areas and levels of the firm. Because of their limited focus on a single or on very few entities and because of the isolated nature of those entities, administrative systems will work best in hierarchic or relational modes.

Type of DBMS system selected

The system type influences the type of data needed, the organizational scope of that data, and the way in which it is accessed. Because of these different focuses, the system type becomes a determinant in the DBMS selection.

Each of the data structural models has different operational and implementation characteristics, and these differences influence the capabilities and restrictions of the data structures themselves. Since the natural structure of the data is one of the major factors in the determination of which DBMS is most appropriate, it is important to understand how each structure looks at data and which kinds of data are most suited for each structure.

Each DBMS allows data to be fragmented, according to the same, or very similar, sets of rules and in roughly the same manner; however, each one employs a different method for connecting these fragments into larger logical data structures.

Each DBMS uses a different type of data structure diagram to represent the connection mechanism which it employs. These diagrams depict (a) the modes of connection and dependencies of the data segments within the larger data aggregates and (b) allowable or supported data access paths between each of the segments. The diagrams represent the structural model for that DBMS and also represent how these data aggregates are defined to the DBMS itself.

Hierarchic structure

The hierarchic diagram presents the data fragments in an inverted tree structure. This inverted tree represents the data segmentation, the segment connections, and the inherent dependencies of those segments. Each tree structure represents the collection of data about one type of entity and is also called a *logical data record.* There can only be one hierarchic structure (Figure 16.2) per database.

A special implementation of the tree structure allows multiple tree structures to be combined into a larger tree, which is also called a logical data record; however, each component tree is still defined as a separate database.

All access to the logical data record is through the base or root segment. It is this segment which contains the unique identifiers, or keys, for the entity occurrence being described. A database contains multiple occurrences of that hierarchy: one for each unique entity occurrence about which data are stored. Each unique entity occurrence within a given database may have its own configuration of occur-

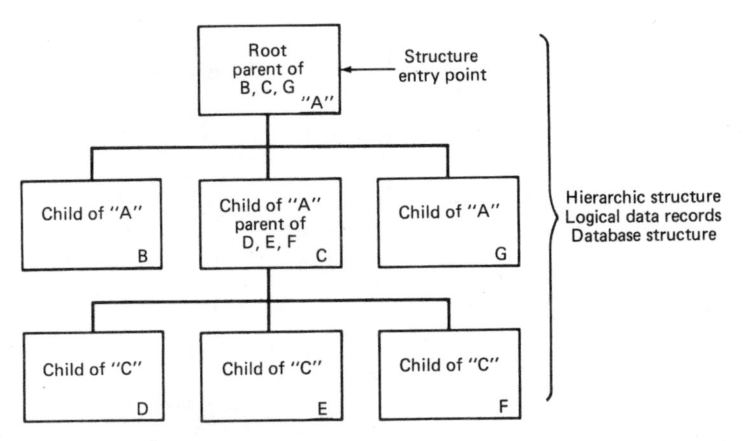

Figure 16.2 Hierarchic structure.

rences or nonoccurrences of each segment type defined with the general entity hierarchy.

The tree structure diagram normally depicts each segment type only once. However, aside from the root segment of each structure, there can be multiple occurrences of any given dependent segment type within the structure.

Each data segment beneath the root segment describes some aspect of the base entity. These dependent segments may be keyed or unkeyed depending upon their contents, usage, and number of occurrences. Dependent segment keys may be unique or duplicated, both within and across occurrences.

Within the hierarchy, root level segments relate only to segments directly dependent on them. The access path (Figure 16.3) to any segment beneath the root segment must include all its immediate hierarchic predecessors or parents on a direct path from the root, as defined in the hierarchic structure.

Segments at the same level below the root (a) cannot relate to each other, (b) must relate as children to only one parent segment at the next higher level within the hierarchy, or to the root itself, and (c) may relate as a parent to any number of child segments, each of which must be one level lower in the hierarchy. These level-to-level or parent-to-child dependencies imply that the lower level segments (children) have no meaning and indeed cannot exist without the higher (in terms of position within the hierarchy) level segments (parents).

The hierarchic model is most effective when (a) each hierarchic structure contains data about a single entity and each entity is relatively homogeneous, having few distinct subtypes, (b) the primary access to each hierarchy is via the identifier of the entity, (c) the entity

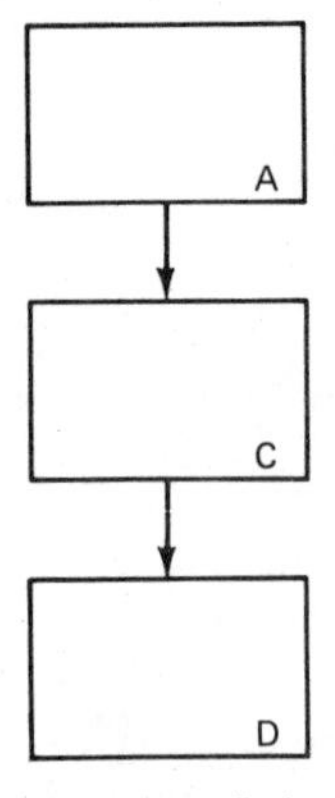

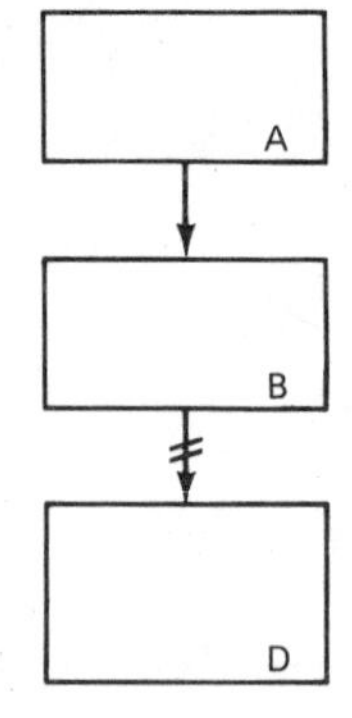

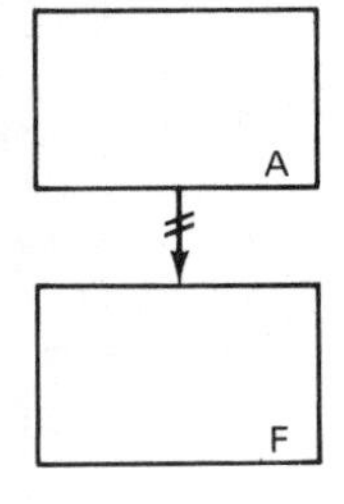

Figure 16.3 Hierarchic access paths.

being described is rich in descriptive attributes and these attributes occur in multiples or not at all, (d) the entity is complete in and of itself, and has few, if any, relationships between it and any other entities, and (e) entity occurrences are processed one at a time.

Network structure

The network diagram has no implicit hierarchic relationship between the segment types, and in many cases no implicit structure at all, with the record types seemingly placed at random. Record types are grouped in sets of two, one or both of which can in turn be part of another two record type set. Within each set one record type is the owner of the set (or parent) and the other is the member (or child). Each record type of these parent-child (or owner-member) sets may in turn relate to other records as either a parent or a child. See Figure 16.4 for an illustration of a network structure.

Each record within the overall diagrammatic structure must be either an owner or a member of a set within the structure. Normally, each set is accessed through the owner record type, which contains the identifier, or key, of a specific occurrence of the set type. Each record type may keyed or unkeyed, and key fields may contain unique or duplicated values. Because each record type in each set can be joined to any other record type in any other set, this structure allows highly flexible access pathing through the various record types for processing purposes.

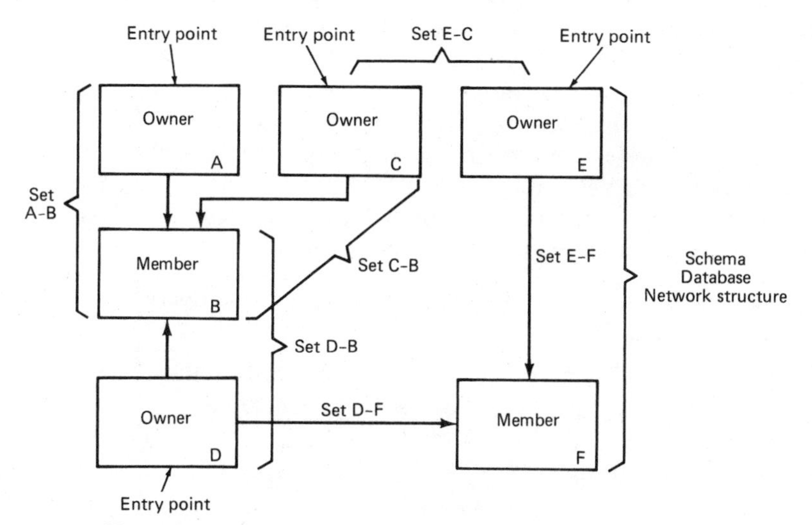

Figure 16.4 Network structure.

Since each record can be related to one or more other records, a network diagram depicts a web of data records with their various interconnections. This composite of all records and all relationships is said to be a schema, and the entire schema is considered to be the database. Firms using network-based DBMSs usually have all data defined under one master schema. A schema may have multiple entry points, one at each owner of each set.

Unlike the hierarchic model where each tree structure is a logical data record, there are no discernible logical data records within a network diagram. Instead each application can create its own logical data records from any combination of sets. These logical data records can then be segregated from the master schema by means of a subschema definition.

There are no levels within the network model, and thus no level-to-level or parent-to-child dependencies beyond those of owner to member. Access to any segment may be direct or through its owner. Any given segment may own or be owned by (be related to) any number of other segments, provided that (a) any pair of segments thus connected must be related through a uniquely named set and (b) any given segment occurrence within a set may have one and only one owner.

Within the network model, hierarchic relationships may be depicted by having one segment own (through multiple sets) many other segments. Each of these owned segments (members) may in turn own a number of other segment, again through named sets.

Within this "hierarchy," however, and subject to set construction restrictions, segment types at the same "level" may relate to each other as either owners or members of sets, and segments at any "level" may relate directly to segments at any other "level."

Segment types may (a) be directly related to segments at a number of other "levels," either above or below the segments' immediate "level," (b) be related to segments outside the "hierarchy" through named sets, and (c) have several hierarchic parents and several hierarchic dependents or children. See Figure 16.5 for an illustration of hierarchic representation in a network data structure.

The network model is most effective when

1. Used to contain multiple entities which are connected in complex interrelationships.
2. The multiple primary accesses to the data structure may be through the identifiers of the entities themselves or through their relationships with other entities.
3. There are multiple entity subsets, each with different attribute descriptors; the dependent attributes of these entities occur in multiples, or not at all.

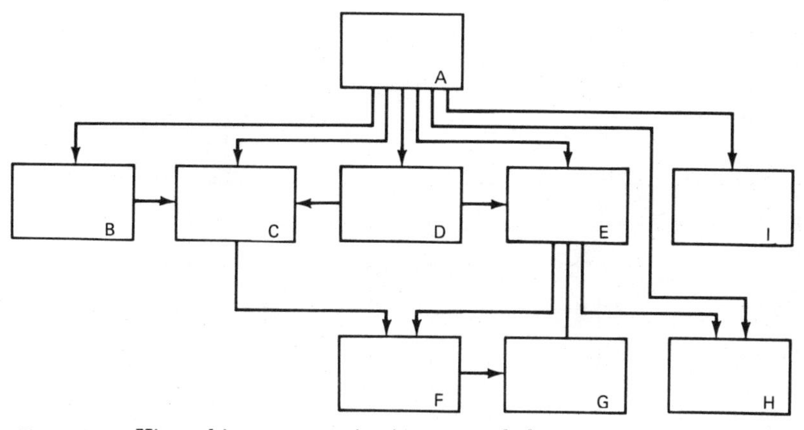

Figure 16.5 Hierarchic representation in a network data structure.

4. There are few if any hierarchic relationships between the entity attributes.
5. The applications need to see the universe of data entities and their relationships, and the process transactions which are aimed at many interrelated entities.

Relational structure

The relational diagram represents each record type in tabular form, and all records of the same type are contained in a single table. The relational model has no implicit structure aside from the table. There are no fixed parent-child nor other relationships within a relational environment. Instead, each table may be related to any or all of the other tables in any number of ways. Any single table, or any combination of two or more tables, may be accessed by any application. Each freestanding table is known as a relation, and each entry within each table is known as a tuple or, more commonly, as a row. Figure 16.6 shows a relational data structure.

Whereas within the network and hierarchic models all data manipulation operations are "record at a time" (where each record must be accessed through its key or through its relationship with, or proximity to, another record), data manipulation within the relational model is "set (or table) at a time" where the set of rows accessed may be as few as one or as many as exist in the table.

Each application can create its own logical data records from any combination of tables, or portions of tables, as needed. These logical data records can then be accessed and manipulated by means of user views or "projections" (Figure 16.7).

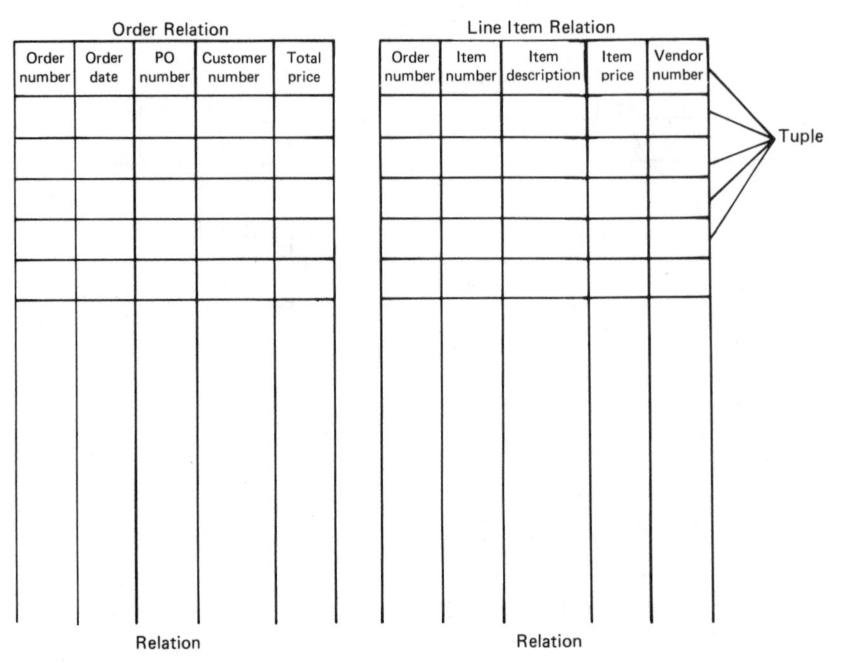

Figure 16.6 Relational data structure.

Although each data table contains data assigned according to a primary key for the entries of the table, there is no explicit sequence to the entries of a table, and the table may be accessed via the contents of any field (or column) of data with it. A table may contain any number of rows, and within a table a row may contain any number of data elements, with the restriction that all data elements must be atomic; that is, they must be defined at their lowest possible level, they must be nonrepeating, and within a table they must be uniquely named.

Each table must have a column of data elements defined as a primary key, and the primary key field of each row must contain a unique value. All occurrences of a data element within a given column of a table must be identically defined and must contain a data value or a null entry.

Tables may be related, or joined, in any sequence (Figure 16.8); any table may start the sequence of joins; and a table may be joined to itself. Any number of tables may be related together, provided that each table of each pair of tables to be joined has a column of data which is identical in definition and is populated from an identical range of values (or domain). Any given table may be directly joined with only two other tables at any one time (A to B and B to C) and the tables must be connected in sequence (A to B, B to C, C to D, etc.), although the sequence of table joins (ABCD, CDBA, DCBA, etc.) is immaterial.

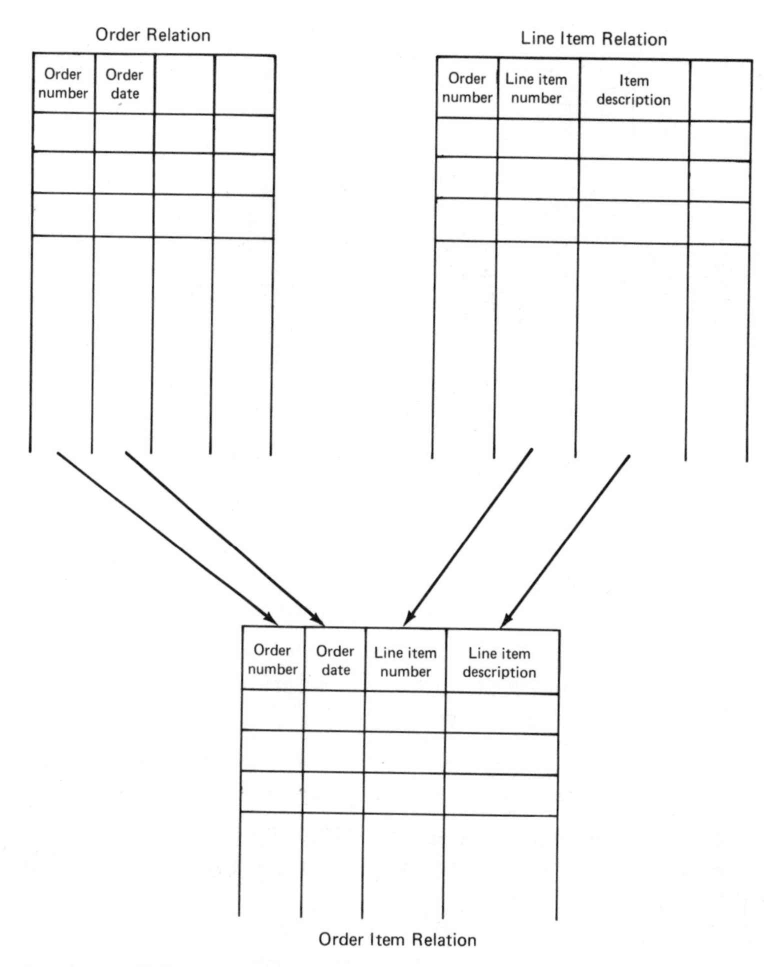

Figure 16.7 Relational projection.

The relational model is most effective when

1. Each table contains data about a single entity.
2. Each entity is homogeneous with little if any differences between subtypes.
3. All data elements within a given table relate to and only to the primary key of the table.
4. The table data need to be accessed in multiple sequences, or via the contents of any data element within the table.
5. There are multiple entities or entity subsets, each with different attribute descriptors and all of which are related to each other in some way.

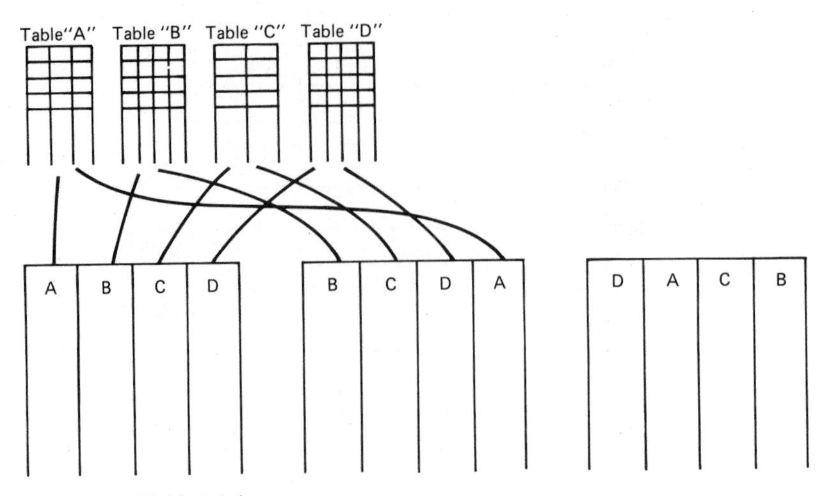

Figure 16.8 Table joining sequences.

6. The dependent attributes of each entity occurs singularly, or not at all.
7. There are no intervening or hierarchic relationships between the entity and each of its attributes.
8. The applications need to see the universe of data entities related in complex ways for retrieval purposes.
9. The applications process updates transactions which can be applied to one entity table at a time and which do not apply to more than one table.

Traditional Analytical Methodology

When using the traditional methods (Figure 16.9), the analyst concentrated on identification of user area processes and tasks, and from those processes and tasks, extrapolated the data necessary to support them. In many cases, the first step in the analysis process was to identify the reports that the user wanted, determine the processing which must occur to produce those reports, and then identify the data elements which must be input to support those processes. This could be equated to a back-to-front process.

The data analysis tasks consisted of identifying the points within the processing sequence at which the input and output files must be placed, and as the processes were analyzed, adding the input data elements to those files which were required for producing the desired output, or which had to be saved from one processing cycle to the next.

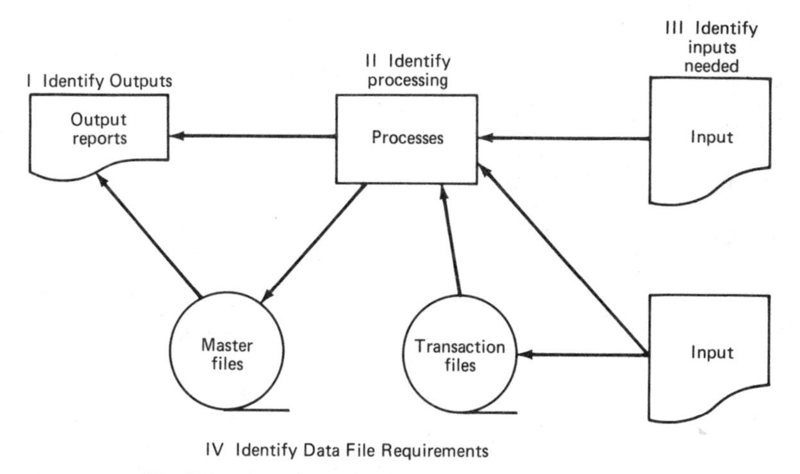

Figure 16.9 Traditional analytical methodology.

Database Environment Methodology

Application development using DBMS technology, with its broader scope and the common data usage environment which it engenders, requires the additional steps of function identification and description and data analysis, tasks which identify the functional, process, and data entities and their relationships to each other. These entities form the basis of the physical data interactions and usage across the firm.

In the database environment, data analysis is that process by which the data structures, data relationships, data dependencies, and data pathing (user views) are developed (Figure 16.10). This analysis develops the information necessary to design the physical database structures. Once the determination of the structural design and requirements of the user's data are completed, the processes may be developed to acquire, and/or manipulate, that data. Whereas in the traditional mode, the processing is the core from which the input and output files are designed, in the database mode the data are the core from which the processing steps are designed.

User Views

We have discussed the need for developing user views of the data. At this point we should provide a definition of a user view.

A definition

A *user view* (of data) consists of the data elements needed for the user to perform a specific function, such as process a transaction. The user view incorporates the data elements which are received, the reference

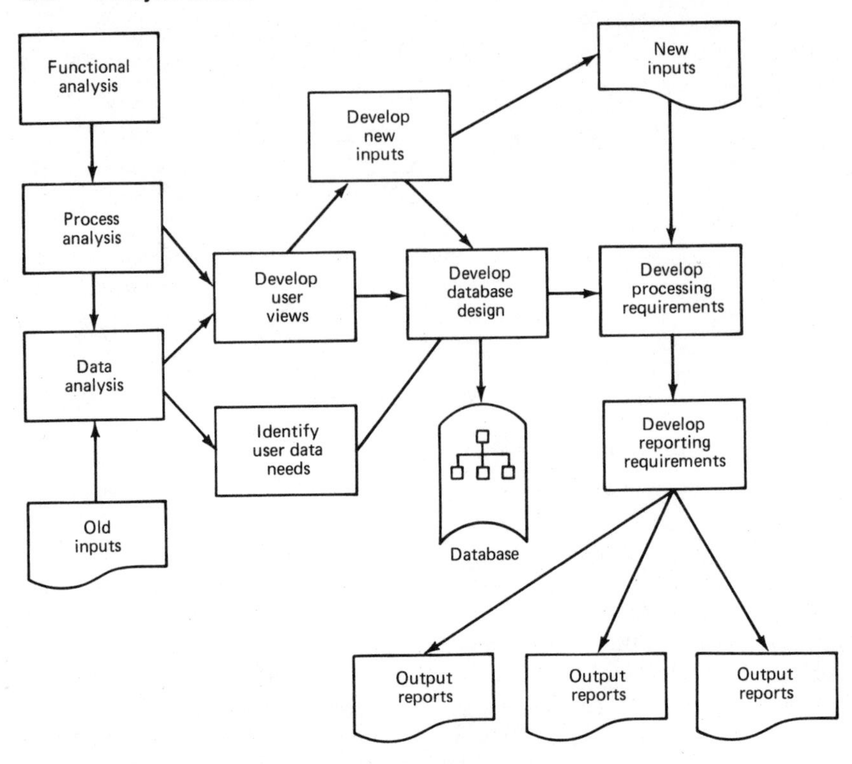

Figure 16.10 Database environment methodology.

data elements needed to process that input, and the data elements which are finally stored in the files of the firm.

At each point in the sequence of a specific process, we need to know the primary identifiers or keys of the input data, the identifiers or data elements which will be used to search the reference files, and the identifiers under which the data will be finally stored. If the data are input in a fragmented form, or if the reference data are retrieved in a fragmented form, or if the data are stored in a fragmented form, the analyst needs to identify the sequence and content of each fragment. Normally, the input, retrieved, or stored data will be more or less coherent. That is it will all be related to a common order, employee, vendor, etc.

User views are usually developed on a transaction-by-transaction basis, although they can also be developed for query and report processing. They detail what data are needed, in what sequence they are needed, and what data identifiers (keys) are relevant. They also identify what processing is performed on the data, in terms of editing, validation, verification, processing, and storage.

User views are needed to identify the basic data element requirements of the users. They are also needed to determine what data are needed in conjunction with other data and to identify when those data are needed. Each user view should be supported by a transaction, a query screen, a report, or a combination of all three.

User views are developed as a result of process and task analysis, transaction and report analysis, and source and usage analysis. Each user view should be supportable by the data file design and by the processing sequences which have been developed. Each user view should be capable of being walked through the data file design, and all data and identifiers should be present in the user view "path" when needed.

Database Methodology and the Entity-Relationship Approach

Designing integrated applications differs in distinct ways from designing processes of more traditional single-user applications. The major differences are in the development of a central design framework; integration of data and its processing, orientation, and sequence of steps; and the inclusion of business entity, business entity relationships, and business entity data analysis.

The design of integrated applications relies heavily on models as one of its integration mechanisms. One of the primary modeling techniques is that of the entity-relationship approach; the others include data flow diagrams and hierarchic process diagrams as well. For instance, the entity-relationship approach refocuses the analyst on the interaction of the functional, processing, and data entities of the firm, their relationships to each other, and their physical characteristics or attributes.

Using this approach, the analyst can build the design around the identification, analysis, description, and interaction (relationships) of these real-world functions, processes, and data entities of the business.

Effective use of the entity-relationship approach, however, requires clear definitions of the terminology and a clear understanding of the concepts involved. The requirement for a clear understanding of the environment is as applicable in the analytical phases as it is in the development of the entity-relationship models; the hesitancy on the part of analysts to develop "models" has hindered the more widespread use of the entity-relationship approach as an analytical tool.

Since the data entities are real, in that they physically exist, they are readily observable. Being real, they interact or relate to each other to accomplish the business of the firm, and they are also relatively stable, and more important, visible and readily describable. They are, in fact, some of the most stable objects in the business environment, more so than business functions or even business processes.

This business entity data analysis looks at the sources of the firm's data and follows the flow of that data through the various business functions and processing steps. Within each function and processing step, the uses and modifications to that original data are examined and documented.

This source and usage analysis results in the identification of the major data entities of the firm, the attributes of those entities which are of interest to the firm, and the natural business relationships which exist between those entities.

The data are arranged into records, and the records are arranged into a logical database structure. The structural logic of the data is then transformed into the physical implementation of the database (physical schema). These analytical steps and transformations are accomplished in an orderly and formalized manner, using new tools and techniques aimed at achieving proper data placement, data access pathing, and access key requirements.

This analysis, called *data analysis*, is achieved mainly through the development and use of modeling techniques. These models allow both the developer and the end-user to define a data model of the business or business segment for which the system is being developed.

This data model of the business, its business functions and transactions, or data events, may be used to ensure that the data have been both properly identified and structured and that the appropriate processing of that data is occurring, or will occur.

The development of the methodological documentation deliverables and of the analytical models may occur concurrently; the documentation may precede the modeling effort, or more appropriately the development of the documentation deliverables and the models may be interwoven.

While the relationships are not exact, the phases of the integrated application design process include all traditional application system development methodology phases plus additional phases which have been structured to place the narrative documentation in perspective through the use of a series of models.

In practice, the information developed in the documentation portions of the methodology provides the bulk of the data, background, and analysis for the additional phases in the expanded methodology.

Data Administration

The data administration (DA) function, along with database administration, is one of the most vaguely defined functions. In many organizations data administration is the name applied to the overall database control and support organization (DSCO). Literally its function is the administration, or management, of data. It is usually more involved with the applications development process and business analysis than with the software aspects of a database environment.

The data administrator normally has the responsibility for the development of corporate standards for data naming and for data definition. The responsibility for data definition extends not only to the data processing implementations of data (i.e., programming language names, internal storage representations, etc.) but also to the business implementations as well. That is to say, he or she has the responsibility of ensuring that the business definitions of the data are clear, accurate, and acceptable to the user community as a whole.

For instance, while the definition of employee *date of birth* might be fairly straightforward, the definition of *date of hire*, *date of employment*, or *date of promotion* might not be as clear or easy. It is the responsibility of the data administration staff to edit, research, and in some cases rewrite the definitions until they are clear to all and have the same meaning to all.

The focus of the data administrator is more on design and analysis than on implementation. As one of the controlling functions of the database control and support organization, data administration provides architectural guidance and support. Its medium is the data, its structure, and usage. The tools are the dictionary and its supporting software.

As such it falls on the managerial end of the responsibility continuum, the end-user end of scope continuum, and the analytical end of the role continuum (Figure 16.11). Data administration's focus is on the early to middle stages of the development life cycle.

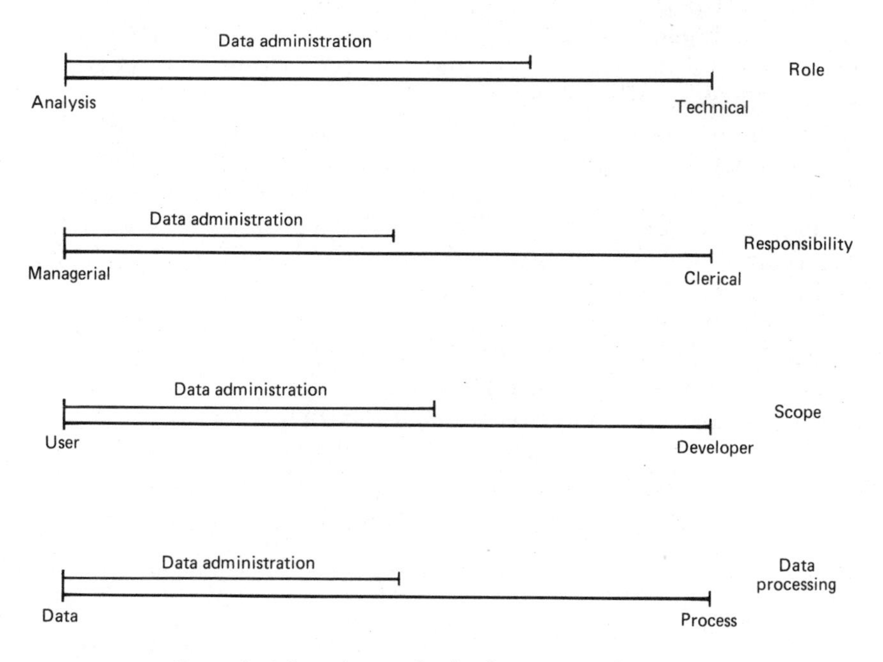

Figure 16.11 Data administration on the development continuums.

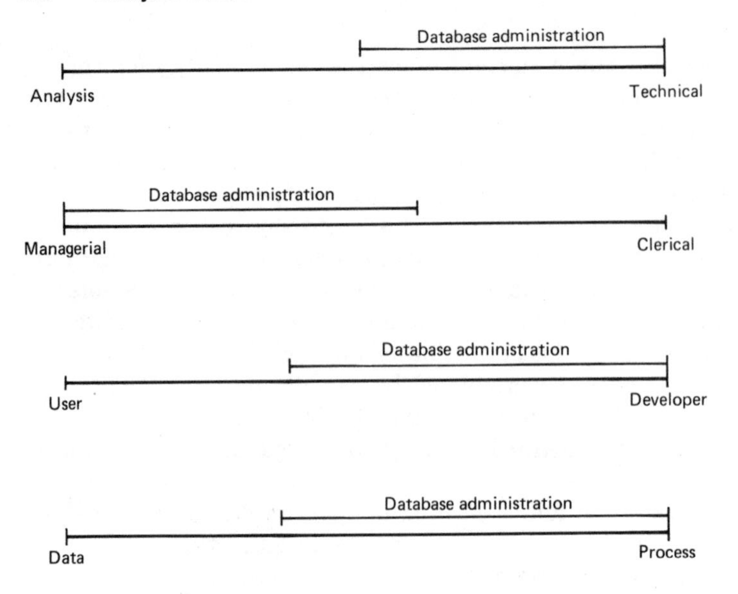

Figure 16.12 Database administration on the development continuums.

Database Administration

If the data administrator is the architect of the database control and support organization, then using the same analogy, the database administrator (DBA) is the general contractor. His or her focus is on software and its implementation, performance, and tuning. With data administration, it is one of the controlling functions and provides technical support and architecture.

Its medium is the DBMS and its access and support software. The tools are the DBMS software and its supporting utilities. The database administrator and staff are among the most technically proficient personnel in the organization, with primary responsibility for the integrity, security, and performance of the physical files of the database. Its function is one of the most critical in the DCSO.

Its profile places it on the technical end of the role continuum, on the managerial end of the responsibility continuum, and on the development end of the scope continuum (Figure 16.12). Database administration focuses on the middle through the final stages of the development life cycle. In addition, it is active through the postimplementation life of the application.

Chapter

17

Microcomputer Systems Implementations

CHAPTER SYNOPSIS

Since many business systems analysis projects must consider the use of microcomputer implementations or microcomputers to augment primarily manual systems, this chapter provides a brief overview of the primary types of products, on a generic basis, which are available for these machines.

Background of Microcomputer Systems

Microcomputer systems are a recent phenomenon within the data processing environment. Compared to mini and mainframe systems they are relatively inexpensive, and using the same base of comparison, they are relatively powerful. The low cost, power, and small size of these machines make them ideal for low-transaction-volume single-user applications, or as desktop executive or middle management workstations.

These machines normally come configured with processor, monochrome or color display, some combination of large-capacity hard disk and smaller-capacity "floppy" disks, and printer.

The addition of a modem to the machine configuration and specialized communications software in both the micro and the mainframe allows the list of micro uses to expand to include

Downloading and uploading of data from the firm's mainframe systems

Access to other micro machines to form a local network with shared data, applications, and peripheral resources

Access to other stores of data and to data provider services such as public databases, ticker and news services, and other information services

Transmission and receipt of data from other sources

Although microcomputers are fast and have a relatively large data storage capacity, they are designed for the end-user. As such, although they have language processors, most users rely on prewritten packages or specialized, general use applications.

Since their introduction in the late 1970s and the early 1980s the quantity, quality, range, and sophistication of the packages available for these small machines have increased almost geometrically. As of this writing there are literally thousands of packages available commercially. These packages cover almost every conceivable application and range in price from a few dollars to a few thousand dollars. Many are available free on public "bulletin boards."

The range of special use applications is too numerous to cover in this work, nor is it our intent to do so; however, it is useful for the analyst to be aware of some of the general purpose packages which are offered. These general purpose packages are easy to use for word processing, accounting (spreadsheets), file creation and update (databases), and graphics applications.

Operating Systems Software

Most microcomputers, as with the mainframe machines, come with some form of operating system or control program, some array of language processors, and communications software. Unlike the mainframes, much of this software is not easily transportable, if at all, between vendors, and many times not even between different models of the same vendor. These products are not, strictly speaking, user application software but are rather application enabling software.

Micro operating systems or control program software

Like their mainframe cousins, these products sit between the user application products and the hardware. They relieve the user of the burden of having to be aware of hardware protocols, machine coding, and physical file creation and manipulation.

These operating systems are tailored to both the vendor and the machine model, although they tend to be very similar in operation across vendors and within vendor model lines. Most operating systems come complete with extensive libraries of utility programs for library maintenance, file manipulation, and hardware and data storage media checkout.

Micro language processors

These products provide users with the capability to write their own applications, either simple or complex, in any number of "higher level" languages. Again, these languages resemble their mainframe cousins and are similar across vendors and within vendor lines.

Micro-mainframe and micro-micro communications

These products, in conjunction with associated communications hardware usually installed within the processor itself, permit the micro to communicate with other micros or with a mainframe. Communications can be either over dedicated lines or via dial-up facilities. These products enable micro networking capabilities, local area networking, and various other machine-to-machine communications facilities. These products do not provide those capabilities themselves but rather enable other software to provide them.

Microcomputer Issues

The issues concerning the choice of micro, mini, or mainframe hardware choice is covered elsewhere in this book. However there are several issues of concern to the analyst which relate specifically to the micro environment.

1. Is the intended activity localized or isolated?
2. Is the volume of data entry low enough to permit user entry?
3. Are there any data security issues which would prevent the application from being implemented on the micro?
4. Is the user sophisticated enough to use the micro?
5. Can the entire application be contained on the micro?
6. Is the growth rate of the application such that it would expand beyond the capacity of the micro in the near future?
7. Are the data on the micro needed elsewhere in the firm?
8. What is the level of internal support for the machines?

9. Can the user make use of the machine when it is not running the intended application?

The Changing Micro Environment

As we stated earlier the personal computer has been available only since the late 1970s and really only since the early 1980s. During the intervening period, we have witnessed an explosive growth not only in the power and capacity of these machines but also in their use by both firms and individuals. While they started as little more than novelty items and offered their owners word processing and spreadsheet capabilities, now many firms implement all but their largest and most data- and communications-intensive applications on them. The increasing proliferation of software packages in all areas of business support will continue as the power and capacity of the machines grow.

This explosive growth and the place of personal computer usage on the growth curves (especially with respect to mainframe systems; see Figure 17.1) is placing a strain on data processing management and on developers.

Because of the rapid increase in the use of these machines and the increased sophistication of the software available for them, it is difficult to stabilize the environment long enough to do a thorough analysis and to make valid software selections.

Additionally, because most firms do not know how to program the machines themselves, most "experts" in the firm are really experts in the use of the various packages rather than in the ability to create these packages in the same manner that mainframe packages are written. This lack of programming skills also precludes most firms

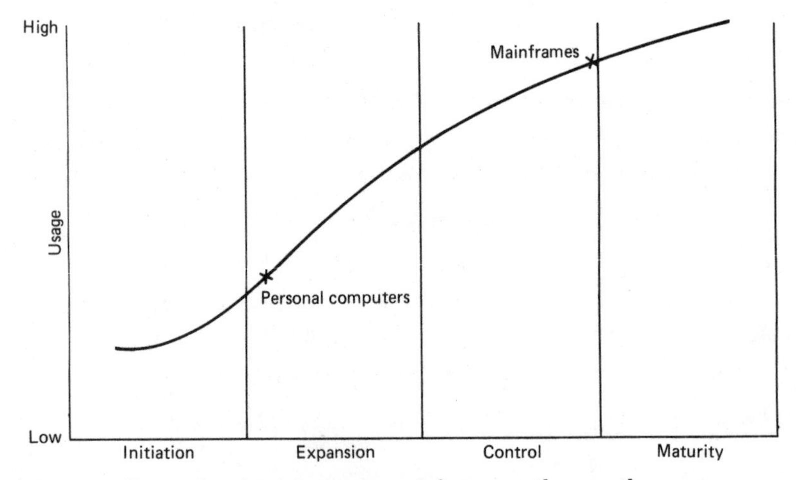

Figure 17.1 Personal computers versus mainframes on the growth curves.

from modifying the packages themselves, and thus they are limited to using those functions and features offered by the package developers.

The lack of standards within the environment and for the machines, the vast variety of manufacturers, and the availability of specialty add-on boards make almost every machine unique. Few firms have sufficient control or sufficient hardware and software standards in place to ensure that an application developed for one machine for one user can even be moved to another machine for the same user, much less to another user's machine. The rapid changes in hardware and the rapid changes in software make even "standard machines" outdated very rapidly.

One last issue of major concern is that of transportability. Although there are widely touted standards, and many products claim compatibility both on a hardware and software level, the analyst should be wary of any mixed system environments. Each package is unique in some respects and tailored to specific hardware and, in some cases, to a model. This uniqueness may prevent movement, not of the package but of the data. This lack of transportability extends even to various versions of a product produced by the same vendor. Great care must be taken to test the products in the proposed environment before incorporating them into any proposals to the user.

The following is an overview of the major types of products currently available for the micro machines. The products chosen are generic, in that they are available for almost all machines and for almost all sizes of machines. The overview presents the general characteristics and capabilities of the product types and is not meant to be a recommendation of any specific vendor or vendor product. Rather it presents alternative environments for the analyst to consider when preparing proposals.

Spreadsheet Packages

In its simplest form a spreadsheet can be viewed as a generalized accounting and numeric analysis package. These packages permit the entry and manipulation of primarily numeric data into a template which is laid out in terms of rows and columns. The number of rows and columns varies from package to package but generally speaking they are sufficient for most user applications.

Each intersection of row and column constitutes a cell, which is identified by a row number (1 to n) and a cell letter (usually a to zz). Each cell can contain one of three general items.

- *Numeric data.* Any combination of digits and special characters (i.e., "." "," "%" "$" "-" etc.)
- *Alphanumeric data.* Any combination of letters, numbers, and spe-

cial characters, generally used as column headings, row titles, or internal explanatory notes or cell identifiers

- *Formulas.* Any combination of cell identifiers, constants, and algebraic operators (i.e., +, -, /, *, (,), etc.), or special function commands (i.e., average, sum, mean, percent, etc.)

From these components the user or the analyst may build any type of simple or complex numeric or financial analysis application. The results of the processing may be displayed on the screen, in printed report form, or in graphic form (Figure 17.2).

The power of these packages allows them to perform complex computations, combined with relative ease of data entry and "programming." Their ability to perform recalculations of the entire spreadsheet facilitates such processes as "what-if" analysis, budgetary planning, cost versus benefit analysis, salary planning, etc.

Word-Processing Packages

Word processors allow the user to use the machine as a text processor, or intelligent typewriter. Text is entered in the normal manner via the keyboard. The power of these packages comes from their text and document manipulation abilities and other built-in facilities, e.g.,

- "Word wrap," which permits continuous typing while the machine takes care of end of line considerations
- Automatic centering
- Pitch and type font changes within the body of the text

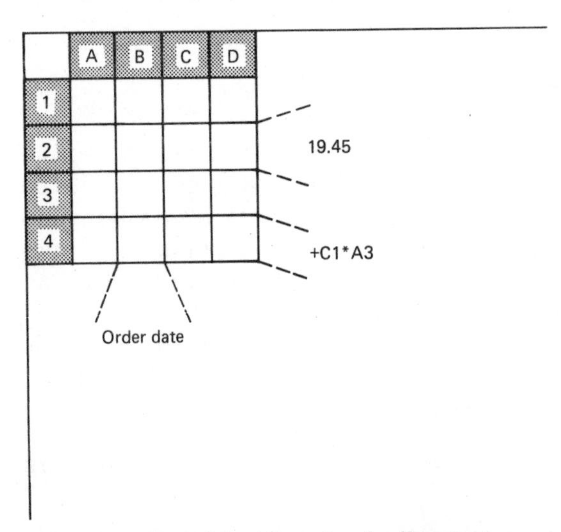

Figure 17.2 Spreadsheet format and cell content.

- The insertion of data, reports, or graphics from other sources within the body of the text
- Automatic pagination
- Automatic headings
- Floating footnotes
- Automatic generation of table of contents
- Spelling checkers
- Bulk text movement and copying features
- Global replacement of single characters or character strings
- Automatic margin justification
- Multiple formats within a single document
- Document format templates which permit simple generation of specialized forms, memos, letters, contracts, manuals, etc.
- Merging of names and addresses from database files into preformatted letters or memos

These packages also permit the documents to be saved to and retrieved from document libraries; they facilitate splitting of one document into several, combining of several documents into one, and moving portions of text from one document to another, etc.

Database Packages

Although they are called "database" packages, these products bear little if any relationship to their mainframe relatives. In essence these are products which allow formatted data files to be created and manipulated. These files may be simple or complex, and may contain as few as one or two data elements per record, or as many as several hundred.

Most of these products are able to easily

- Develop screen forms for data entry and data update
- Sort the files on any field or group of fields
- Select a record or group of records based upon some set of field content criteria
- Produce reports and labels
- Merge and split files
- Create automatically maintained relationships between records in one or more separate files
- Extract all or part of the data within the files for use in other products, such as spreadsheets and word processors

- Load files with data extracted from other products or from data downloaded from the firm's mainframe files

The most common uses for these products are for

- Mailing list creation and maintenance
- Product file creation and maintenance
- Order file creation and maintenance

Electronic Mail

These systems are used for the electronic bulk transmission of reports and memos from one user to many others. In addition, they allow the user to receive and transmit informal notes to other users on the network. These packages are usually used in conjunction with word-processing systems and normally require routing through some intermediate mainframe (micro networking) or the existence of a local area network (LAN) linking the micros together.

Memos, reports, and notes are stored in user-specific "mailboxes" which normally are only accessible by the designated owner. Mailbox contents may be browsed, stored locally, or responded to and rerouted to the transmitter.

Project Management Packages

One of the newer offerings for the micro market are project management products. They combine pieces of three other products: spreadsheets, databases, and graphics.

Project management packages come complete with data entry facilities which permit the user to define the tasks and the personnel availability for his or her project, verification facilities which permit cross-checking task to task, and one or more graphic facilities which permit the generation of project schedule graphics (PERT, CPM, or Gantt). Graphics support may be automatic, i.e.,the package generates the charts from the task list, or manual, i.e., the package allows the user to draw the charts with automated assistance. Once created these charts can be printed or modified at will, segmented into smaller pieces, or merged into a larger schedule.

Many packages come with project templates, that is, predefined task lists which the user may use as is, add to, or modify as needed. Many packages also come equipped with predefined reports and data collection or data entry forms.

Computer Graphics

Most micro products come complete with graphics capability. This is especially true for spreadsheet products. Since the primary method of data input or data display from a micro is via its monitor, most products come equipped with graphics of some sort. These graphics may be as simple as a preformatted input screen or the graphics displays may be very complex.

Because of the increasingly fine resolution of the monitors themselves, many packages have been created to allow the development of presentation-quality pictures. Some packages come complete with predrawn figures which the user can expand, contract, move around on the screen, combine with other figures, and color using an infinite variety of colors from the graphics "palettes." Many of the newer products allow picture sequences to be displayed, in effect allowing the user to "animate" the presentations. Most of these products allow the screen-generated pictures to be printed on printers in addition to being displayed on the screen.

Chapter

18

Controls, Auditing, and Security Analysis

CHAPTER SYNOPSIS

Security of data, physical plant, and processing controls are the major areas which affect business systems. They may make the difference between acceptable and unacceptable systems proposals, and between successful and unsuccessful systems. Additionally, any development project analysis must examine current security procedures and determine the need for any enhancements to them.

This chapter discusses the concepts of security and controls, the areas where they are most needed, the determination of security and control requirements, and their impact on the final system proposal.

Security

The topics of security, auditing, and controls analysis are usually overlooked during the systems analysis process. The logical relationship between the three topics is sufficiently tight that they can be treated in one unit; however, they are also different in many respects. Security is the more all-inclusive topic, with auditing and controls being the tools used to ensure that security is maintained. Many areas of security have already been treated in previous chapters, and much of the information gathered during the functional, process, and data analysis phases applies to the evaluation of security requirements of the firm and of specific applications.

Although security should be an ongoing concern for the firm, most firms do not address these issues until the data processing environment has migrated to the third stage of the data processing growth curve and often it is well into the third stage. Thus an analyst working within a first or second stage environment, or even in an early third stage environment, should pay careful attention to this area of analysis (Figure 18.1), since it is in these early stages that security is most likely to be minimal or lacking altogether.

The analyst should also remember that the firm may not be in the same place on the growth cycle in all areas (especially in the personal computer area), and thus each security analysis should be undertaken with the particular growth stage of the particular user and implementation environment in mind.

What Is Security?

Security relates to protecting the firm's records and resources from unauthorized access, modification, or other interference. It includes an analysis of ownership, access, modification, and use, as well as a determination of what protective or restrictive measures must be taken to ensure adequate protection of the firm's files. Whereas the bulk of systems analysis concentrates on determining which processing activities are being done and which should be done, security analysis concentrates on who should or should not be doing those processing activities and when those activities should or should not be done. This examination should not necessarily be carried out from the perspective of functional needs, although they obviously play a critical part in this area of analysis.

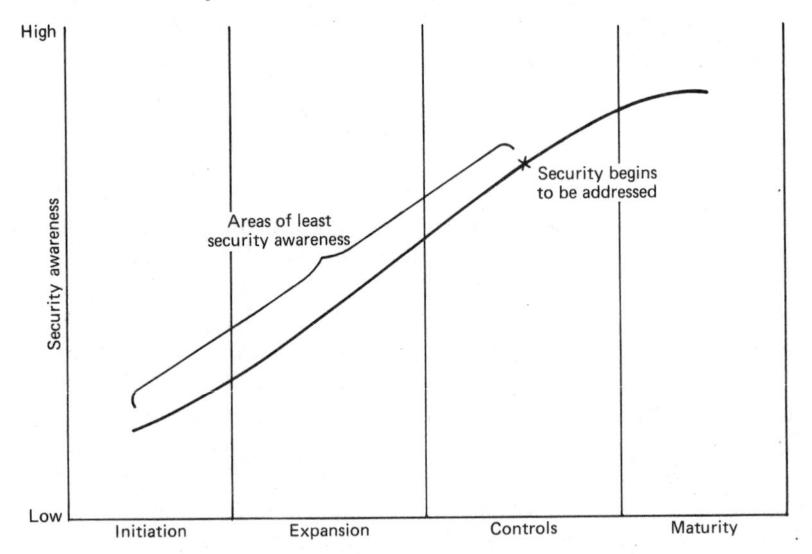

Figure 18.1 Security awareness and the growth cycle.

We tend to think of security in terms of preventing physical entry; however, by extension it can also be viewed as preventing logical entry as well. In the business environment this determines who should be allowed to access, modify, and remove business records. It is immaterial whether these records are physical or automated. Security is also concerned with record retention and safekeeping. Generally any analysis of security requirements involves answering a series of questions relating to "who can do what with what."

Securing the firm's data and information resources may also entail examining the physical resources which allow access to those data. These resources include printer facilities, data storage (both machine readable and hardcopy), and data access facilities (the most obvious of these being terminals and personal computers).

Generally speaking, security can be said to include (see Figure 18.2)

1. Those procedures which ensure that the correct people perform the proper actions at the proper times and use the proper resources to do so
2. Those procedures which ensure that those persons who are not authorized to perform certain actions or use certain resources are restricted or otherwise prohibited from doing so

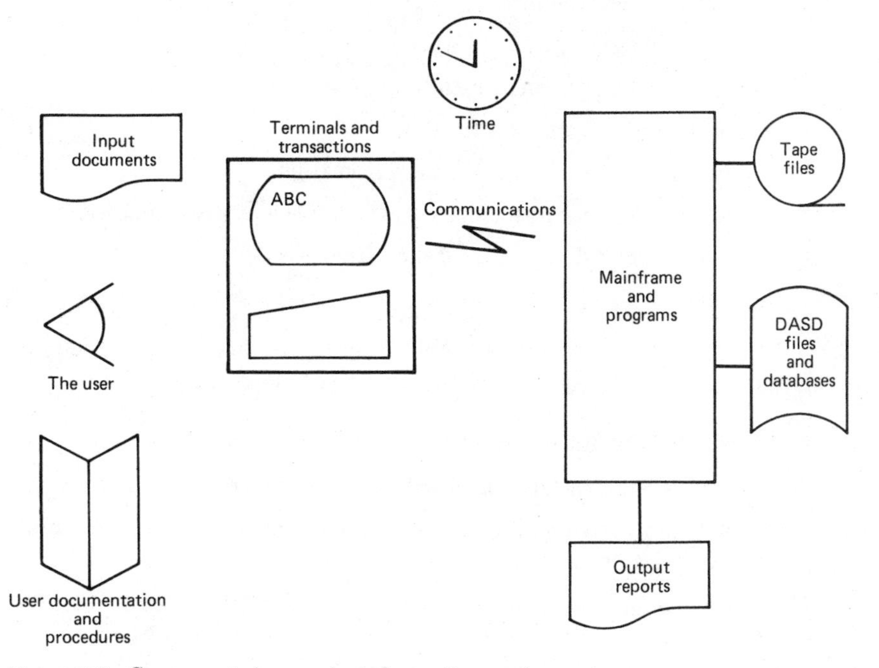

Figure 18.2 Components in a review of security requirements.

3. Those procedures which ensure that authorized persons do not perform authorized actions or use authorized resources at unauthorized times, or for unauthorized reasons
4. Those procedures which detect and record any violations of the above three items

Security Analysis

Security analysis thus involves making a determination of what must be done, when it must be done, how it must be done, what is needed to do it, and who should be doing it. Security analysis also includes an examination of the physical access points to data.

Security analysis usually begins with an evaluation of securable resources. This is done on the assumption that the more sensitive, valuable, or irreplaceable the resource, the greater the effort which should be expended securing it.

The first set of questions thus involves the material to be secured. Generally this involves an inventory of the firm's files and the records within those files. Included in this inventory should be

1. All original source documents
2. All internal documents and forms
3. All reports for both internal and external use
4. All manual and automated files
5. All copies of all manual and automated files
6. All manuals, product descriptions, schematics, drawings, etc.
7. All memoranda and correspondence files

This inventory should be as complete as possible, and most of the information should already be available from the environmental analysis documentation. Each item in the inventory should be annotated as to

1. The source of the material
2. The official owner of the material (not necessarily the source)
3. A list of those persons or units who are authorized to access, modify, and otherwise use the material
4. The time or conditions under which it can be destroyed or thrown away
5. The name of the person or unit that is authorized to destroy it

In some cases there may be more than one person or unit involved in each of the above; in this case, all names should be listed.

Once such an inventory has been completed, the next step is an examination of each of these files and records to determine if they need to be secured. Not everything within the files of the firm needs to be secured, and security may not be needed all the time. For instance, quarterly financial reports need to be secured before publication but not afterward. Likewise, prior to publication, manuscripts should be secured from nonemployees, but not after.

This determination of security needs should be done on an item-by-item basis and should evaluate

1. Is the material available from other sources or is it wholly generated within the firm?
2. Is the material protected by license, patent, copyright, nondisclosure agreement, contract, or other legal mechanism? If so, does the license, patent, etc., belong to the company or to someone or some organization outside the firm?
3. Does the firm have any existing method of classifying documents, files, and records? If so, does this item fall within that security program?
4. If no such program exists, does the analyst need to devise such a program for the user area? The depth and extent of the security requirements are dependent upon the user area management and the nature of the work done by the unit. Not all areas of the firm have security requirements. In some cases the security need may be minimal, and in others, such as product development, personnel and payroll, etc., it may be extensive.
5. Is the content of the material such that it warrants security? For instance, employee payroll lists with names and salaries are usually considered very sensitive information; however, statistical reports on payroll are normally not sensitive. Reports containing product costs or specifications are usually sensitive unless such information is otherwise publicly available.
6. Is the nature of the information on the document such that its disclosure would be damaging, embarrassing, or otherwise detrimental to the company?
7. Is the nature of the material such that the company is legally precluded from disclosing it, such as employee medical information?
8. Is the nature of the material such that disclosure of it to unauthorized persons might place the firm in a position where it could be open to a lawsuit?

The answers to these questions should provide the analyst and the user manager with enough information to determine and classify those documents which need to be secured. The classification scheme can be as simple or as complex as is necessary to obtain the desired level of security awareness.

Once the classification process is complete, the analyst needs to determine how to achieve the desired level of security. Since most of the firm's resources, especially its data resources, should be used in conjunction with legitimate, that is, authorized, activities, the analyst should next cross-reference the documents and resources to the activities and procedures which use them and make a determination as to whether those uses agree with the security profiles established for the documents. For instance,

1. Are the persons who perform the activities with those documents authorized to do so?
2. Are the activities being performed in the correct and established time frames?
3. Are the activities being performed in the documented and authorized manner?

For each activity which includes a secured document, have procedures been established to ensure that the security of the documents are both known and maintained? Are there procedures established to ensure that document security classification is clearly established? Have all procedures which use secured documents been modified to ensure that they include actions which ensure that the security of the documents is maintained when they are not actively being used?

The actions to be taken with respect to these questions may include any or all of the following:

1. A set of rubber stamps
2. Nondisclosure agreements or other similar documents
3. Locking file cabinets
4. Restricted copy and distribution procedures for certain material
5. Restricted use of copier and other duplicating facilities
6. Sign-out–sign-in procedures
7. On-premises vaults
8. Off-premises storage

Physical Access Security

It is not enough just to secure the data. The analyst must also determine whether access to the terminals which provide access to the data has also been secured. Are the terminals in a physically secured place? Do the terminals have locking mechanisms? Does the telecommunications system provide facilities for sign-on and password security facilities? Have these facilities been implemented? Are they enforced?

In many firms employees are assigned sign-on codes and passwords. Many times these codes and passwords are simple and easy to remember. In many cases these codes are written on the terminals themselves or in plain view. For those applications which require access security, these items should be examined carefully and more stringent procedures should be initiated where necessary.

Personal Computer Security

Because of the great ease with which data can be loaded into or unloaded from these machines and because of the extreme portability and small size of floppy disks, data security becomes a grave issue in those environments where user systems are developed for the personal computer.

The issue becomes even more critical where these machines have the ability to download or upload data to and from a mainframe. Although most of the software for personal computers is copy protected by the vendors, the simplicity of the data access methods offered by the machines allows almost anyone with even a limited amount of knowledge to read and thus copy data. The copy protection mechanism of the software does not extend to the files themselves. Anyone with a version of the baseline software can access any user file.

Some packages permit the data files to be secured, but here again, the protection is often rudimentary at best. Although encryption of data (the scrambling and encoding of the data so that it is unreadable without the decryption mechanism or code key) is possible, many firms do not insist upon it, and the difficulty of using these techniques is such that many users dispense with it altogether even where it is available.

The ability of many personal computer users to bypass the protection mechanisms of the vendor software packages, and the ease with which they do so, makes any data security mechanism seem almost transparent.

Aside from these issues, a firm's lack of standards and its inability to enforce those standards compound the problems of personal computer data security.

Controls

Unlike security analysis, which seeks to determine the importance of the firm's files and to determine what protective measures, if any, must be applied, controls analysis seeks to determine what measures are needed to ensure that the protective measures are being followed and that the necessary work of the firm is completed accurately. Controls serve two functions: (a) to ensure that security mechanisms are being followed and (b) to ensure that the data and resources being secured are correct and accurate in the first place. Whereas security analysis and its resulting measures are administrative in nature, controls analysis is mainly an accounting function.

Controls are used to ensure that

1. All documents in a group of documents have been processed correctly
2. Generated reports are numerically accurate
3. All steps in a predefined series have been taken and completed satisfactorily
4. Physical counts match calculated counts
5. Certain actions are taken at the prescribed time
6. Actual costs are kept within budget
7. Information passed from workstation to workstation is accurate
8. Staffing counts are kept within predefined levels
9. Actual actions match planned actions
10. Actual results match planned results

In all of the above cases, the goal is to ensure that what is actually done matches expectations and that the information upon which actions are predicated is accurate.

Normally, controls are applied to actions which

1. Are very complex
2. Involve a large number of steps
3. Involve a large number of items
4. Involve high cost
5. Are highly susceptible to error
6. Require a large number of people or units to complete

Controls are very similar to quality assurance tests in that the intent is to ensure that all actions have been taken correctly. Normally controls are built into the actual processing steps of an activity; they usually involve verification of manual or automated actions by persons other than those who performed the original actions.

Controls may be applied item by item, such as rechecking order or invoice line item extensions and totals, or they may be applied to groups of documents together, such as generating a grand total of order totals and checking them against a manually generated total.

In our discussion of batch processing, we talked of batch controls (Figure 18.3). These are control totals, which may be in the form of total dollars, total items, total documents, or some other procedure. These totals are accumulated before the data are processed and are compared to the totals from the machine-generated reports. If the totals agree, the batch is assumed to be in balance; if they do not agree, the individual items are examined to determine whether an item was missed or entered incorrectly. When the error is found, the item is corrected and the entire batch is then resubmitted for processing.

Totals are also generated across batches to ensure that all batches are entered. Since most systems generate large reports, financial or otherwise, most reports have one or more levels of totals. They may be totals of numeric columns, total items, etc. If the report is exceptionally large, there may be several levels of subtotals within the body of the report. These are usually at logical breaks in the data and are usu-

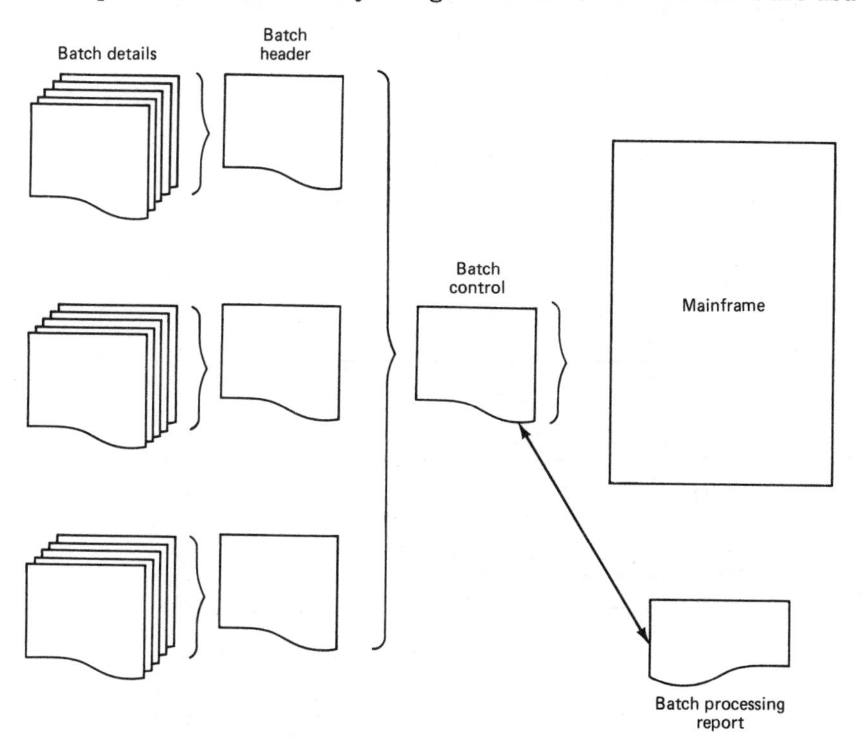

Figure 18.3 Batch control checking hierarchy.

ally taken at some change in the major report sequence field. Aside from providing information to the report recipient, they also provide a mechanism for checking the report's accuracy and completeness, either on a manual or automated basis.

This type of totaling procedure is also used when files are processed to ensure that the proper number of transactions of each type have been processed, and that records have not been either added to or deleted from the file erroneously.

Controls may also take the form of checklists and signature sheets, where certain information is recorded as steps are completed. This might include start and stop times, item counts, operator or clerk initials, batch numbers, machine meter readings, verifier or checker initials, supervisor initials, etc. Figure 18.4 is a sample batch header form.

Order Entry
Batch control header

Date ____________

Time ____________

Batch number ____________

Order number	Number of line items	Total price	Total quantity
____	____	____	____
____	____	____	____
____	____	____	____
____	____	____	____
____	____	____	____
____	____	____	____
____	____	____	____
____	____	____	____
____	____	____	____
____	____	____	____
Control totals	____	____	____

Entered by ____________

Verified by ____________

Figure 18.4 Batch header form (sample).

The analyst, working with the user and with the work and process flow analysis documents, as well as with the documents themselves, must determine where errors can originate and must devise procedures which can prevent the errors from occurring or detect those errors that do occur.

In other instances, the analyst may have to determine methods for controlling the processes themselves. This may entail the creation of control logs or carrier documents which will accompany the documents being processed. Figure 18.5 shows a sample batch control form.

The process of determining which controls to impose is very similar to a checks and balances operation. For instance, controls should ensure that the people who perform a certain activity are not the same ones who check it or that testing or sampling procedures are devised which check work in progress. In some cases, separate subsystems or

Order Entry
Batch control

Date ______________

Time ______________

Batch number	Total lines	Total price	Total quantity	
______	______	______	______	
______	______	______	______	
______	______	______	______	
______	______	______	______	
______	______	______	______	
______	______	______	______	
______	______	______	______	
______	______	______	______	
______	______	______	______	
______	______	______	______	
______	______	______	______	Control totals

Written by ______________

Verified by ______________

Figure 18.5 Batch control form (sample).

activities may have to be devised for ensuring that proper controls are in place. These controls could entail summary reports, comparison reports, variance reports, exception reports, statistical analysis, or in extreme cases duplication of processing.

Controls may also take the form of "audit trails" (Figure 18.6), which are reports or files generated from the business transactions of the firm. Each transaction and activity is recorded as it is processed. Each file add, change, and delete, each file access, each program execution, and each document processed is recorded or logged. This ensures that if there were a need to re-create a portion of the work, there would be sufficient information, i.e., a "trail," to do so. It also ensures that if there were a need to determine whether an activity was performed, who performed the activity, or when the activity was performed, there would be sufficient information available to make such a determination.

Most, if not all, DBMS products offer optional logging of transactions to the databases which they maintain. Many make this type of

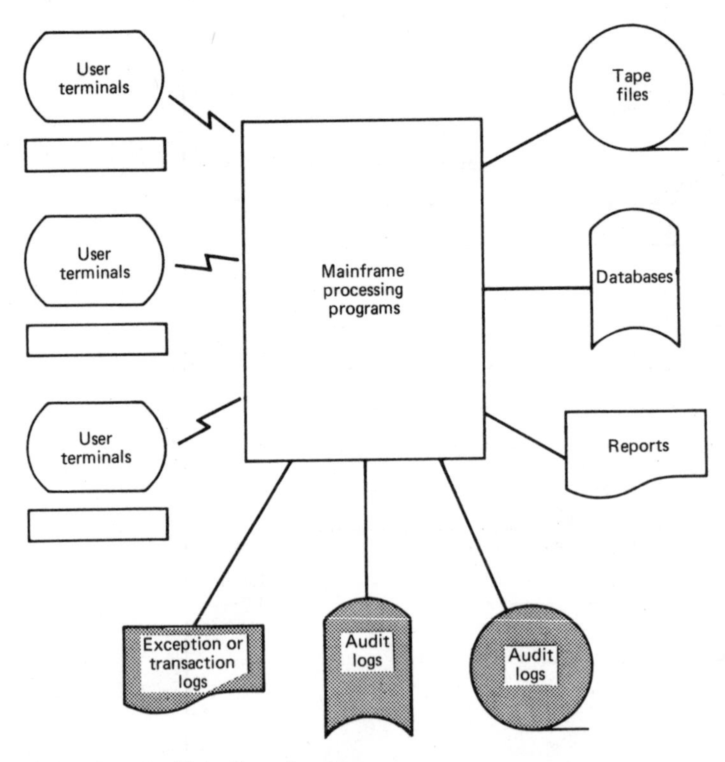

Figure 18.6 Audit trail mechanisms.

logging mandatory, in that the DBMS will not operate without logging files. Most telecommunications systems also offer logging of the online transactions processed by them, as well as logs of all file accesses and changes. Most products offer security facilities, and access violations are logged to these files as well.

The type and extent of the controls to be imposed will often depend upon the nature and sensitivity of the work being performed, on the ability to reproduce the work should it prove to be defective or in error, or on the impact on the firm if any error should continue to be undetected.

To illustrate, it may be acceptable for errors to appear on internal documents, but totally unacceptable for the same type of error to appear on a document which is to be sent to the customer. Again, it may be acceptable for costs to be slightly in error when work is being done internally (such as is common with internal data processing chargeback systems) but unacceptable when pricing an item for sale to a customer.

Aside from the above reasons for establishing control mechanisms, one of the major reasons for controls is improvement of the auditability of the firm's operations by both internal and external accounting functions. These types of controls are usually applied to all activities

1. Where cash, checks, or other monetary instruments are involved
2. Which are reflected on the firm's books, balance sheets, or profit-and-loss statements
3. Where company employees are handling customer funds
4. Where real loss to the firm could occur
5. Where the firm could incur liability if errors are not detected
6. Where assets or resources of the firm which have real value are being handled.
7. That ensure that the security and integrity of the firm's files and records are maintained.

Auditing and Auditability

All firms and organizations, both publicly and privately owned, governmental, charitable, profit making, or not-for-profit, must maintain books of accounts which record all business activities; transactions of a monetary nature; and acquisition, use, and disposal of inventory, raw materials, work in process, machinery, buildings and land, furniture and other supplies, stock, etc. In short all assets and liabilities. Periodically the status of these accounts must be reported to the various governmental entities for purposes of tax assessment, licensing, etc. In the case of corporations, partnerships, etc., stockholders and regulatory bodies must receive reports on the financial status of the firm on a periodic basis.

The books of accounts or other records are normally prepared for the firm either by the management of each user area or by the internal accounting areas, although in some cases outside accounting firms may perform this work. Regardless of who prepares the account books, the work of verifying and validating their accuracy and the related accounting functions is usually performed by external or "public accounting" firms, or by certified public accountants (CPAs) who are charged with the responsibility of making sure that such reporting is complete and accurate. These certifications are required by and are acceptable to the various agencies to whom the reports must be made.

These accountants or auditors use various techniques to perform this verification and validation, in most cases relying on the records which result from the controls imposed by the firm for these purposes. In many cases they will run their own tests, reports, etc., to ensure that their results match those provided by the firm.

If in their examination they determine that the controls or records are unacceptable or insufficient to properly verify or validate the business transactions, the auditors may ask that certain additional controls be imposed to assist them in their activities. These controls may take the form of new reports or additions to existing reports. Auditors may also impose requirements on the firm to retain original records for certain periods of time, to retain copies of critical transactions, to generate duplicates of various files and to undertake similar activities, and the analysts may be required to assist the auditors in this work. The analyst must be sure that any statements of requirements include the system verification requirements of the auditors.

The Concept of Time

Time is a vital concept which must be kept in mind during the analysis process, more specifically the changes which occur over time, time dependencies, and the concept that certain things while changing over time may need to be retained as of a certain point in time. The passage of time implies changes in state and in conditions. Things which are valid now may not be valid a few moments from now. These effects of time are incorporated into the data which are stored as a result of business activity and are also incorporated into the things which trigger certain business activities.

Examples of the importance of time are

1. The price of a product is stored in a file along with other descriptive information about the product. This file could be a product catalog, or it could be a data processing file. When an order is placed, it may be for a specific product at today's price, at the price when it is shipped, at a price to be determined, or at a price which existed yesterday. Which price is associated with that product line item on the order depends upon what the business rules state or upon the agreements made between the sales department and the customer.
2. An invoice is prepared and sent to the customer. It states an invoice total and may or may not reflect specific terms, such as "2/10 net 30," which means that if the invoice is paid within 10 days, the customer may deduct a discount of 2 percent; however, from the eleventh day on the terms are the net of the invoice.

   That same invoice, once mailed, begins to age. The firm needs certain follow-up procedures, at the 10-day point, the 30-day point, the 60-day point (when the invoice is past due), at the 90-day point, etc. Payments against the invoice may be received at any time after invoice mailing. That payment may reflect discounted price, the net price, or the net price plus penalty charges (if any) for late payments. The firm must be able to determine what payment is expected, at what times, and how to evaluate payments received. At each of these points different actions need to be taken, all of which depend upon remembering two points in time: invoicing date and today's date.

3. Employee reviews, salary actions, project schedules, and delivery schedules all require different actions depending upon the dates involved.

Time must be factored into data storage considerations, processing considerations, and most schedules. These time equations must be incorporated into how long it takes to perform an action, how much time can be expected to pass before a response is forthcoming, and when the final action must be taken.

The analyst must determine just how critical the time factor is in the user area and to the user work schedule. The time factor can be measured in terms of seconds, minutes, hours, days, weeks, months, etc. Some schedules are fixed, others are highly flexible. The clerk on the phone with the customer needs information instantaneously, but the correspondence secretary may have information response times of hours and perform in a perfectly satisfactory manner.

Another area where time is relevant is in the controls and audit trails of the firm. Each transaction should be stamped with the date and time as it enters the firm. All changes to the books and records of the firm should also be dated to reflect when the change was made. All reports and documents should be dated, and in some cases time stamped to indicate when they were created or received into the firm.

In some cases only the item itself need be dated; in others it may be necessary to date the entire record with "from" and "to" effective dates and to store those records in historical files or archive files.

The analyst should determine the existence of, and the need for, any historical files or archiving of records as the firm's procedures and activities are analyzed. In many cases new history files may be needed or the procedures for updating old ones may need to be revised. In all cases the analyst should pay attention to these filing systems and to the material stored in them.

Some of the questions which should be asked when analyzing existing filing and archiving procedures are

1. When was this file established?
2. Under what conditions is it used?
3. How often is it used?
4. Under what conditions and how often is material deleted from the files?
5. Of what use is the information in the file?
6. Does the filing system have an index or reference system? Can information be easily obtained from the files?
7. Is the index or reference system kept up-to-date?
8. Is the need for this information still valid?
9. Does the information in this file exist elsewhere, and can this information be duplicated easily?
10. What controls exist to track material added to or removed from the files?
11. Can the material be easily identified as to the time period when it originated?

Part

5

Applications

Laws of Project Management

I. *No major project is ever installed on time, within budget, with the same staff that started it. Yours will not be the first.*

II. *Projects progress quickly until they become 90 percent complete; they then remain at 90 percent complete forever.*

III. *One advantage of fuzzy project objectives is that they let you avoid the embarrassment of estimating the corresponding costs.*

IV. *When things are going well, something will go wrong.*
- *When things just can't get any worse, they will.*
- *When things appear to be going better, you have overlooked something.*

V. *If project content is allowed to change freely, the rate of change will exceed the rate of progress.*

VI. *No system is ever completely debugged: Attempts to debug a system inevitably introduce new bugs that are even harder to find.*

VII. *A carelessly planned project will take three times longer to complete than expected; a carefully planned project will take only twice as long.*

VIII. *Project teams detest progress reporting because it vividly mainfests their lack of progress.*

Anon

Chapter

19

Basic Types of Systems Projects

CHAPTER SYNOPSIS

This chapter presents a brief discussion of the most common types of applications found in most firms, and the cases which follow (Chapter 20) illustrate the environments and issues facing firms in various types of industries. The application discussions and the cases illustrate the variety of implementations as they appear in different types of firms.

The Typical Business Flow

In order to understand the general characteristics of the most commonly addressed applications, it is important to understand the typical business flow (see Figure 19.1).

Most organizational activity can be traced to some origination in the marketing and sales functions of the firm. These superfunctions include the functions of

- Product development
- Market research
- Sales
- Sales analysis
- Sales management

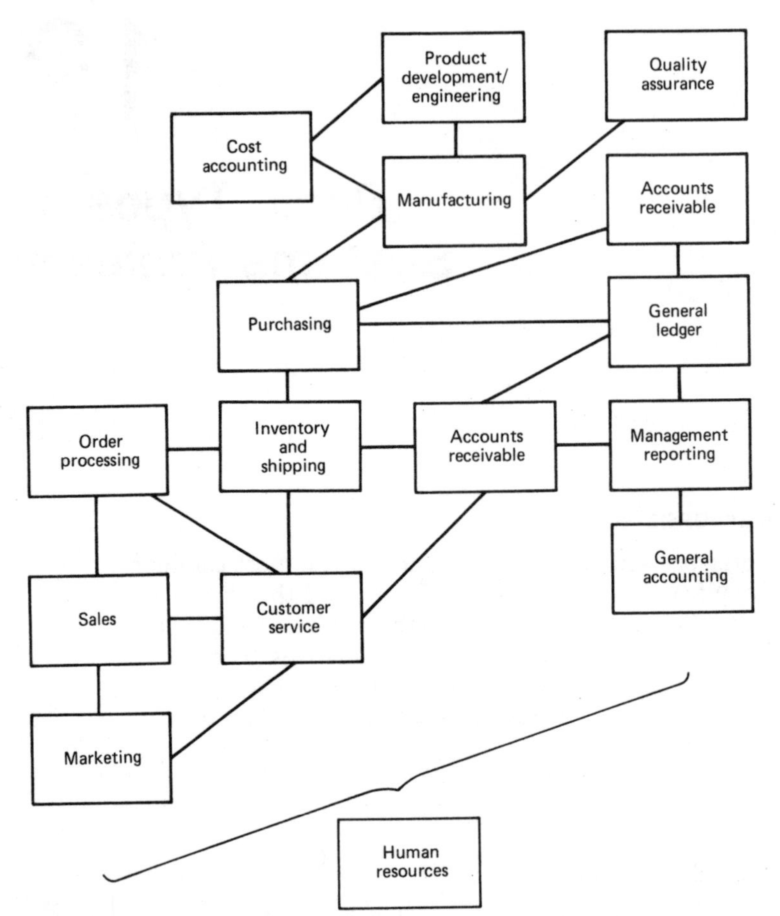

Figure 19.1 Simplified functional interactions in manufacturing.

It is here that customers are identified and sales are made. These sales are received by the firm in the form of orders for the firm's products and/or services. The incoming orders are edited, validated, credit checked, and placed in the firm's processing stream.

If the orders are for products, the products are taken from inventory, shipped, and invoiced. As product is taken from inventory, the inventory records are updated and the quantities on hand are then checked against minimum or optimum stock levels. If the quantities have dropped below these levels, orders are placed for replacement stock.

Once invoiced, the orders become part of the firm's receivables accounts, or moneys due. It is for these satisfied orders that bills and statements are processed, and against which payments are received.

The replacement of inventory products, either by purchase or by manufacture, causes the firm to place orders with other firms. These orders become part of the firm's payable accounts, or moneys owed.

When replacement stock is needed, purchase or manufacturing orders are cut, vendors for stock or materials are contacted and orders placed, receiving orders are prepared, and inspection areas are notified.

In some cases the firm's business calls for it to receive and manage moneys for its customers. The transactions associated with this activity are recorded in a separate set of accounts, called customer accounts (in banks, called demand deposit accounts, savings accounts, etc.). When the firm's business calls for it to receive and hold a customer's property, accounts are also set up to record each physical asset transaction.

Transactions associated with financial aspects of the firm's business, both internal and for customers—moneys due, owed, received, paid out, retained as profit, and property received, disbursed, and retained—are recorded in a structured set of individually titled financial records called general ledgers.

The firm must also recruit, hire, and manage people as employees of the company, develop and maintain records on those employees, and compensate those employees for services rendered. The recording of these records and of compensation information is handled by a complex set of functions known as human resources (or by their individual function titles: personnel and payroll).

On a periodic basis, the firm and its auditors must prepare financial reports, perform inventory counts, and issue financial statements to various agencies and to the firm's owners.

It is important for the analyst to be familiar with the general aspects of each of these types of functions and the data and processing which are associated with them. It is equally important that the analyst recognize that although most companies have similar functions, applications, and systems, they are not identical from company to company, or even from company division to company division. Their differences have been shaped by the individual company's

- Business
- Industry group
- Age
- Size
- Complexity
- Organizational structure
- Managerial and operational philosophy

- Culture
- Particular accounting structure
- Data processing organization
- Sophistication in terms of its use of data processing
- Reliance on data processing services to run its business
- Systems development budget

These factors, singly or in combination, result in the requirement that each analysis and development project make no assumptions with respect to the processing required, steps within those processes, or the data which drive the project. Each project must be assumed to be new and unique.

The industry descriptions highlight the differences in basic company functions and processes; however, across industries many similar functions exist as well. Given their similarity across industries, it should not be surprising that these functions are either administrative or managerial in nature, although they are all greatly affected by the particular type of business of the firm and by the operational systems which service them.

In order to understand the functional differences, the analyst should first have an understanding of each of these basic functions. Although there are many functions within the typical organization, there is a limited set which appear in some form in every organization, whether the firm engages in direct product manufacture, acts as agent for the sale of products made by others, or delivers a service.

Among these functions are

1. Order processing
2. Inventory
3. Customer service
4. Accounts payable and accounts receivable
5. General ledger
6. Marketing and sales
7. Payroll and personnel (human resources)

The functional descriptions which follow are general overviews intended to familiarize the analyst with the terminology and basic activities and processes of these common functions. The descriptions are, of necessity, very general and incomplete but do reflect many of a firm's functional activities, the general flow of information, and the complexity and importance of the function as well as its relationships and interactions with and impact on other functional areas of the firm.

Order Processing (Figures 19.2 to 19.4)

All companies receive and process orders for the products they produce or for the services which they offer. These orders may be spontaneously received (such as mail orders) or they may be solicited in some way (either through a sales force, by direct mail advertising, or by media coupon advertising).

Each order must have certain basic items of information, including: who placed the order, what was being ordered, and how many were being ordered. The orders may also contain item prices, applicable taxes, and shipping and handling charges. Unless the firm is using a

Figure 19.2 Order processing flow.

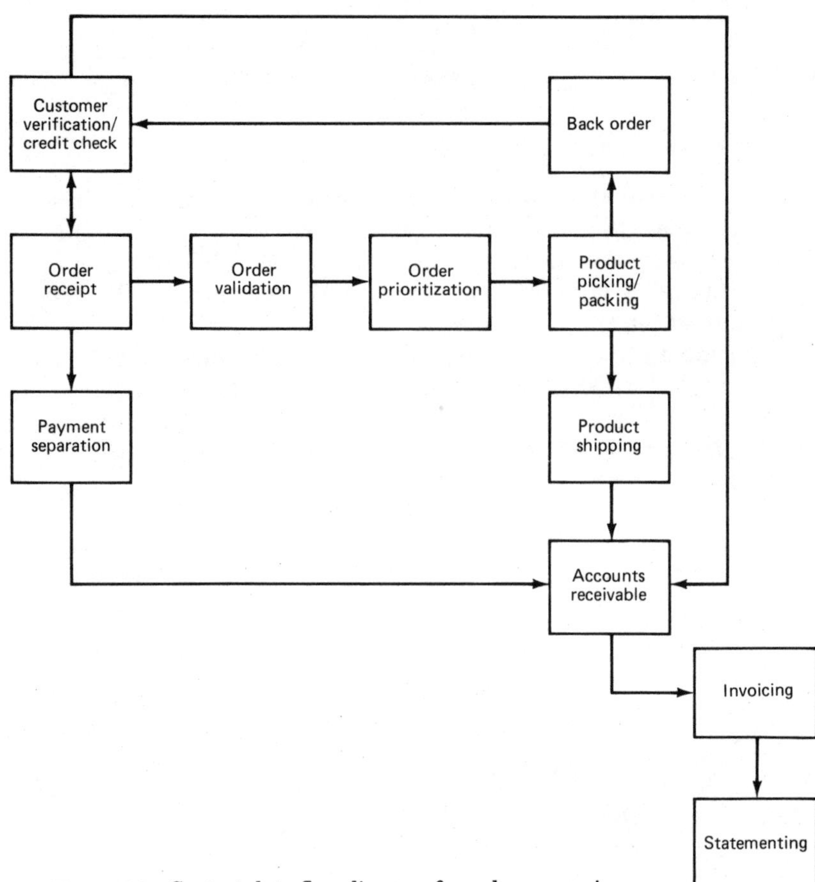

Figure 19.3 Context data flow diagram for order processing.

sales staff or a preprinted order form, these orders will usually be free form, less than complete, and may contain inaccuracies and discrepancies. In some cases, the item ordered may not even be one the company sells. In others, it may be for items which the company no longer offers. The item description may be complete and accurate or may be vague and incomplete.

Orders may be accompanied by payment which might be complete and accurate, overpaid, or underpaid. In many cases the orders may be cash on delivery (COD), may have a credit card number, or may ask for later billing.

Each order must be examined, edited, corrected where necessary, and entered into the pending order files of the firm for later satisfaction. This edit and correction process usually includes verification of item description, price, discounts taken or offered, taxes, all additional fees and charges, and all extensions and totals, as well as addition of the company item number where necessary.

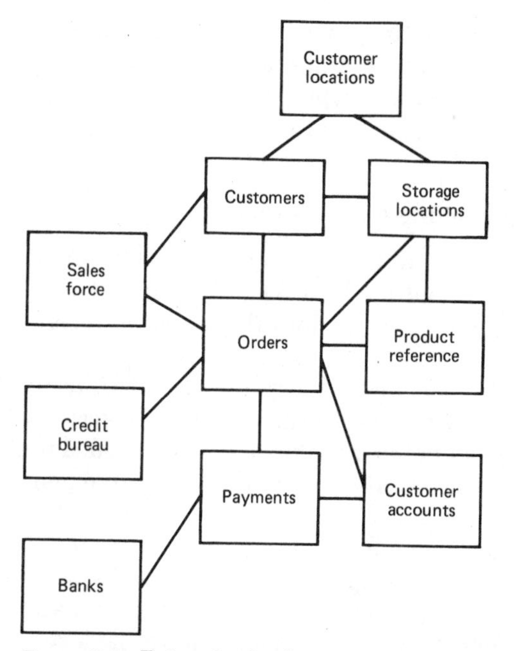

Figure 19.4 Enterprise level entity-relationship model for order processing.

In addition the order must be checked to ensure that the customer name and address are properly entered, that any shipping or billing information is complete and accurate, and that any sales or commission information is recorded for that order. Any priorities associated with the order, or for any item on the order, must be noted, and any complete ship, partial ship, or delayed ship instructions should also be noted.

If the order contains a credit card payment or postshipment billing, the order is usually sent to the credit check area where any credit card authorizations are obtained or a credit check may be done on the customer. If payment was by check, the firm may elect to delay shipment until the check clears the bank.

Once the order has been accepted, the processing of the order varies with the firm, and, it is highly dependent upon whether the order is for goods or services. This processing is industry, company, and product specific, but, generally speaking, it involves one of the following.

1. Picking the product from the storage locations (warehouse bins, shelves, etc.), packing the product for shipment, determining the method and costs of shipment, cutting a shipping document, and actual shipment
2. Cutting a work order for product manufacture, manufacturing the

product, inspecting and packaging the product, packing the product for shipment, determining the method and costs of shipment, cutting a shipping document, and actual shipment

3. Executing the buy, sell, or trade for the customer and issuing a confirmation
4. Delivering the service or arranging for the service to be performed

Once the order has been satisfied, unless it is a cash or COD order, the order, along with any additional charges for shipping and handling, is sent to accounting where it enters into the accounts receivable processing streams.

Many firms open accounts for their customers and record transactions for buys or sells into these accounts. In these cases, the account application contains all billing, shipping, pricing, and other needed information, and the "order" itself need only refer to the base agreement and "override" any default information.

Other firms accept what are known as standing or master orders. Standing orders may call for periodic shipments of products, for shipments of a set number of all new products, or may set the terms and conditions for future orders which may be issued against it. These standing orders are usually placed either (a) to save paperwork on future orders, (b) to take advantage of volume discounts, or (c) to lock in current prices for future deliveries.

In some cases, where the customer has multiple locations, the customer may place master orders which specify the total quantity, price, central billing information, and the quantities that are to be shipped to each location. For some master orders, each location is billed separately and the central office is sent a consolidated statement.

Orders may specify that the product be shipped to the ordering customer's own locations, shipped to client locations, or stored by the company for later shipment.

Inventory (Figures 19.5 to 19.7)

Those firms which offer products for sale and which either manufacture the product themselves, wholesale the product (buy from the manufacture and then sell to retailers), or act as agents or consignment sellers usually sell from inventory.

The inventory may be on premises, in separate warehouses, external to the firm (held by another), or of such a nature that the product itself is stationary and only ownership changes hands (e.g., houses, buildings, land, natural resources, etc.).

Regardless of where the physical product is kept, the products are assets of the firm. Thus their sale must be accounted for, and the firm

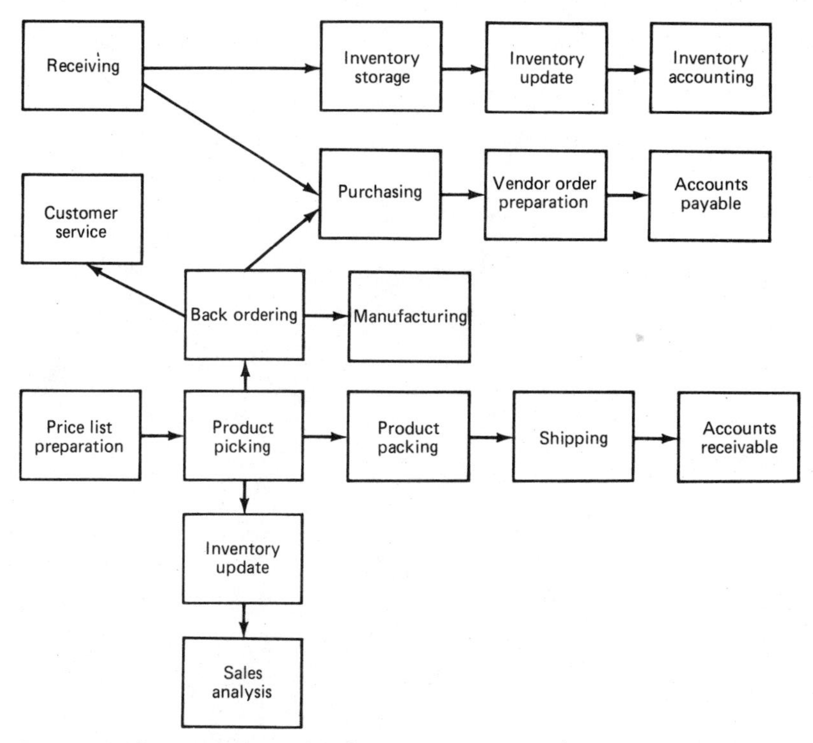

Figure 19.5 Inventory processing flow.

must keep accurate records of transfers into and from stock or of how much is left to sell (in the case of depletable resources).

Each transaction to and from inventory (the total product available for sale) is recorded on an ongoing basis. Periodically the firm will check the recorded quantities on hand against the actual quantities on hand (a physical inventory count). In many cases variance reports are issued and discrepancies accounted for. In all cases the recorded inventory is changed to reflect the actual quantities on hand (since that is actually what is there).

During day-to-day order and shipment processing, the removals from inventory are recorded and the remaining quantity is checked against some predetermined quantity (usually called a "reorder point"). In some cases, stock cannot be replenished (such as one-of-a-kind items); in this case, once the last item is sold, the records are marked inactive.

For those items which can be replenished, when the reorder point is reached, a restock, repurchase, or remanufacture order is issued. These reorder points are calculated by a variety of methods and usually reflect estimated sales volumes, time to restock, "safety limits," manufacturing times, and a number of other variables.

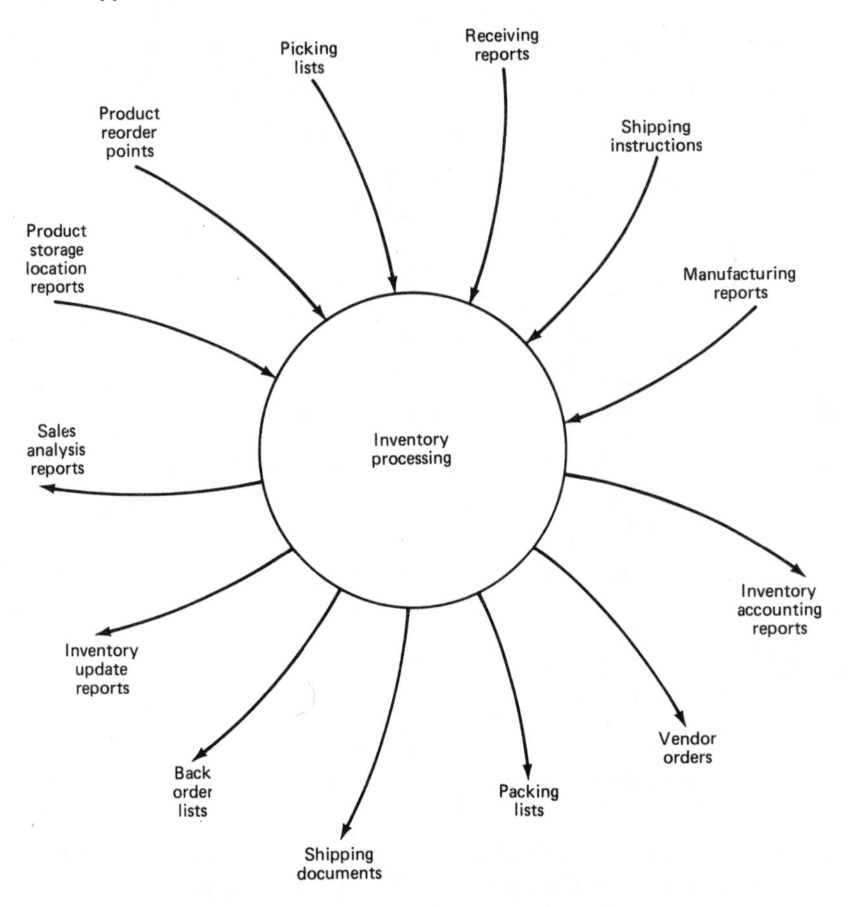

Figure 19.6 Context data flow diagram for inventory processing.

If the firm miscalculates its sales volumes or there are delays in obtaining new stock, orders for depleted items may be "back ordered." Back ordering simply means that the order is set aside and placed in a file, usually by item, until such time as new stock is obtained. In these cases, the customer is normally notified of the back-order condition and given an estimate of when the stock will be available.

In some cases, there may be sufficient stock to partially fill the customer order, in which case (assuming the order and company policy permit it) the available stock is shipped and the remainder is placed on back order.

Manufacturing firms which use raw materials to make their products or which use purchased components which are assembled into the company's product have complex systems in place to ensure that sufficient materials are on hand for manufacturing. These systems are usually called materials requirements planning (MRP) systems, and

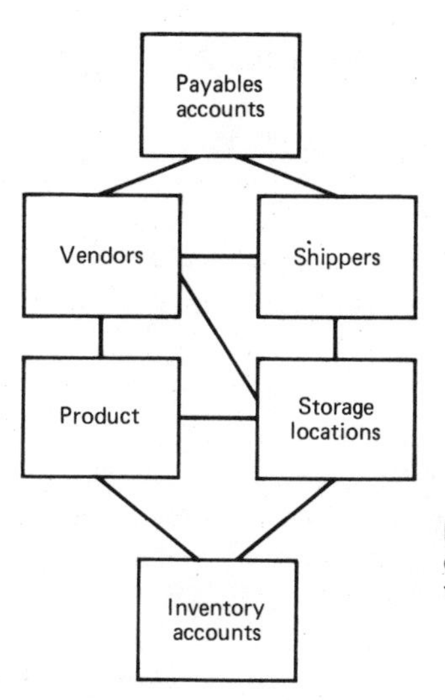

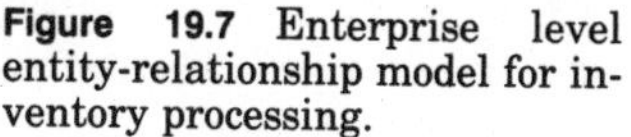
Figure 19.7 Enterprise level entity-relationship model for inventory processing.

work from product bills of materials and work schedules. MRP systems attempt to work backward from the manufacturing processes and use lead times, purchasing times, inspection times, etc., to automatically determine when and how much material must be ordered to ensure that the company's production lines are continually supplied.

Since inventory of finished product, "work-in-process" product (product which is partially manufactured), and raw materials represent an investment of funds, most firms attempt to maintain a balance between having too much stock and not enough stock so that manufacturing will not stop or customer orders go unfilled. Inventory systems work in conjunction with sales analysis systems, manufacturing systems, purchasing systems, and materials management systems to ensure that this balance is maintained. Periodic reports are usually needed to determine slow- or fast-moving items, stock turnover, cost of inventory, and funds planning for new stock purchases.

Some firms buy and sell items for their customers which the customer may either (a) not need possession of, (b) not be able to take possession of, (c) buy and sell so quickly that physical possession is impossible, or (d) not want possession of.

In these cases the firm must maintain accounts which keep track of the customer's inventory of holdings and periodically report to the customer on the status of those holdings and on details of all transactions against inventory. Examples of firms which use this type of inventory processing are brokerage firms (which normally hold their customers' securities), banks (checking and savings accounts), shipping and transportation brokers (where vehicles and/or cargo are constantly moving), and real estate managers (whose inventory is buildings and apartment or office space).

Customer Service (Figures 19.8 and 19.9)

All firms deal with customers of some type. These customers may be individuals, other firms, or governmental units. The firm may deal with the customer directly (over the phone or in person) or indirectly (by mail or other communication media).

These customer dealings, aside from straight sales, usually require either resolving some problem, providing some information, or providing other than normal service to the client. To handle these types of dealings, most firms maintain customer information files on (a) all customers or (b) selected customers. These files are used to maintain information on customers; their preferences, idiosyncrasies, likes, and dislikes; special handling requests; or more normally their transaction histories.

These files must be available when the customer calls or communicates for information, with a problem, or with a special request. Firms which extend credit to customers maintain credit information in these files along with transaction and payment histories.

Firms which maintain open accounts for their customers (ongoing accounts where the customer's transactions are recorded) must maintain the details of each transaction against those accounts and be able to tie the accounts to the customer's nonaccount information (i.e., credit, residence, demographic, billing, and shipping information, etc.).

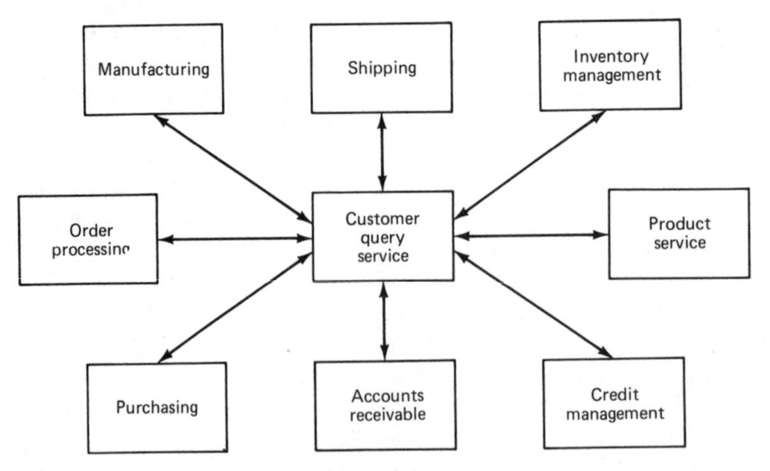

Figure 19.8 Customer service process flow.

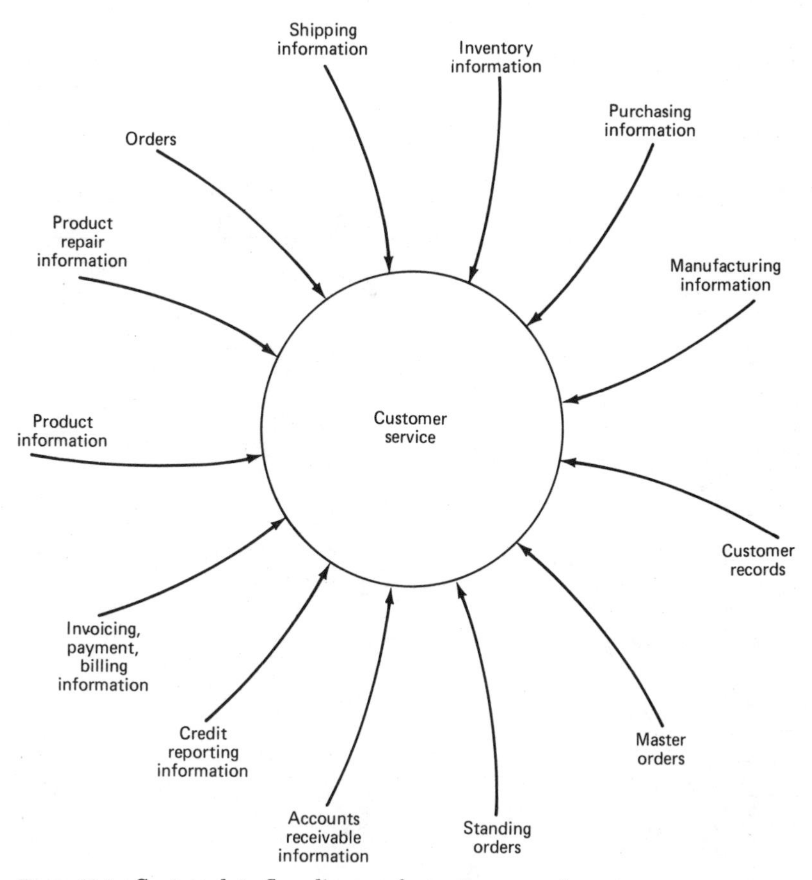

Figure 19.9 Context data flow diagram for customer service.

Accounts Payable and Accounts Receivable (Figures 19.10 to 19.13)

Once a customer order has been processed and either the product shipped or the service rendered, the order is transformed into a receivable item. That is, the order now represents a payment due, which the company now expects to receive.

This order may have been an isolated order or one of many orders from the same customer. The normal sequence is for an invoice to be rendered to the customer (if it wasn't presented upon delivery). That invoice, which looks very similar to and contains many of the items of the original order, details the items covered, the total amount due from the customer (including fees and charges), and payment terms.

Traditionally the customer will have 30 days to pay, although some terms call for payment at month end and others offer discounts for

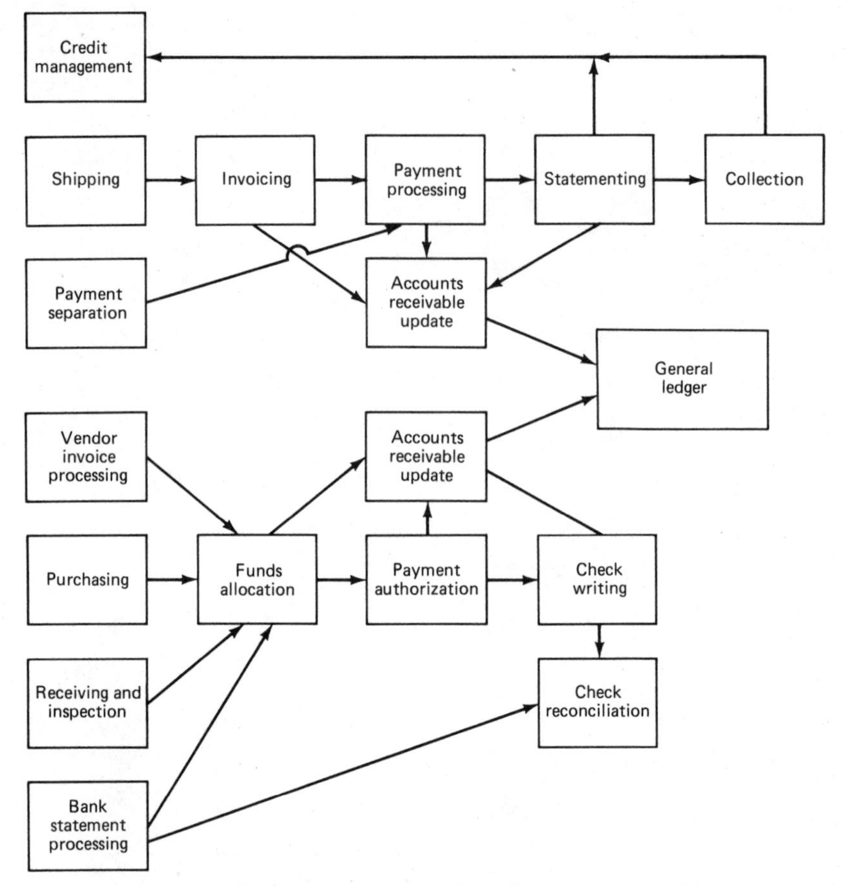

Figure 19.10 Process flow for accounts payable and accounts receivable.

early payment. At the 30-day point, or at month end, all outstanding invoices, along with any payments received, or amounts due or credited to the customer for overpayments or returns, are summarized on a customer statement. These statements normally list all invoices closed since the last statement and all open (unpaid) invoices.

The statement may further break out open invoices into an "aged" listing, beginning with invoices over 90 days old, followed by invoices that are 60 to 90 days old, 30 to 60 days old, and under 30 days old.

Part of the accounts receivable processing includes receipt and application of payments. When making payments, customers are asked to indicate in some way which invoices are being paid. If the customer does so, the payments are applied to the indicated items. If not, they are normally applied to all open items starting with the oldest items first and working back to current items. This cascading effect of pay-

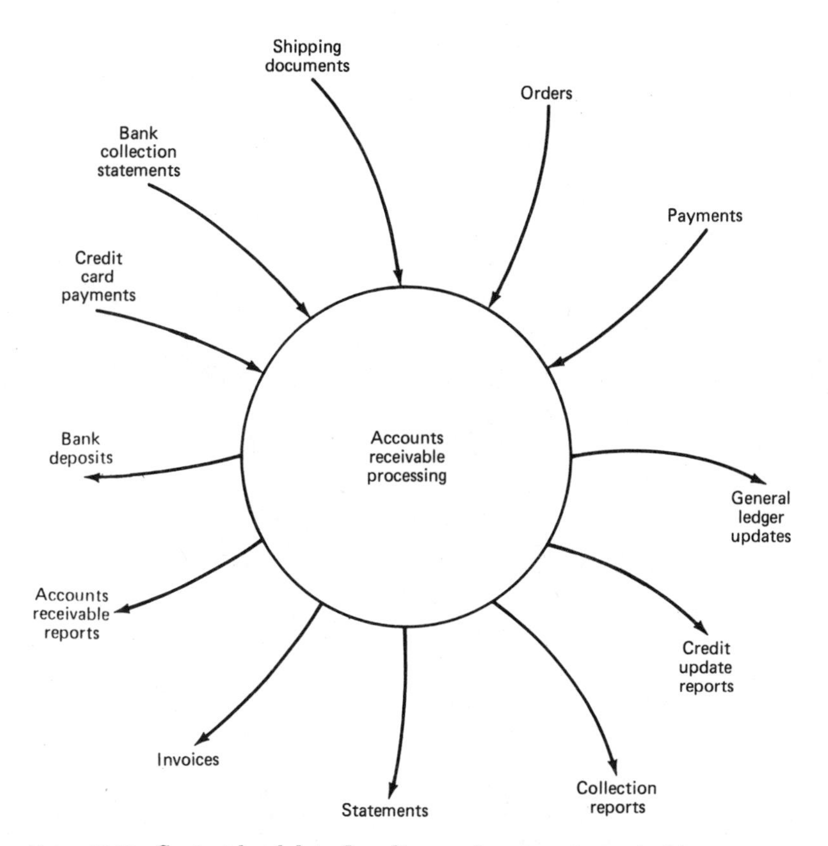

Figure 19.11 Context level data flow diagram for accounts receivable.

ment application usually results in partial payments being applied to some invoices.

All firms which maintain open accounts in this manner also maintain a set of ledger accounts, usually one for each customer where ongoing records of invoice and payment history are recorded. As payments are received and items are closed, the payments are routed to the various other company accounts in any one of a variety of ways and in accordance with how the company's general ledger system has been defined.

All firms purchase supplies, materials, and services from outside firms or individuals. They may be for product manufacture, office needs, or a variety of services. These purchases are transacted by issuing orders to vendors. Once the goods, materials, or services have been received, inspected, or otherwise accepted, the firm will receive the vendor's invoice, which becomes a payable item for the company. That is, it represents moneys which it owes to someone else. The same

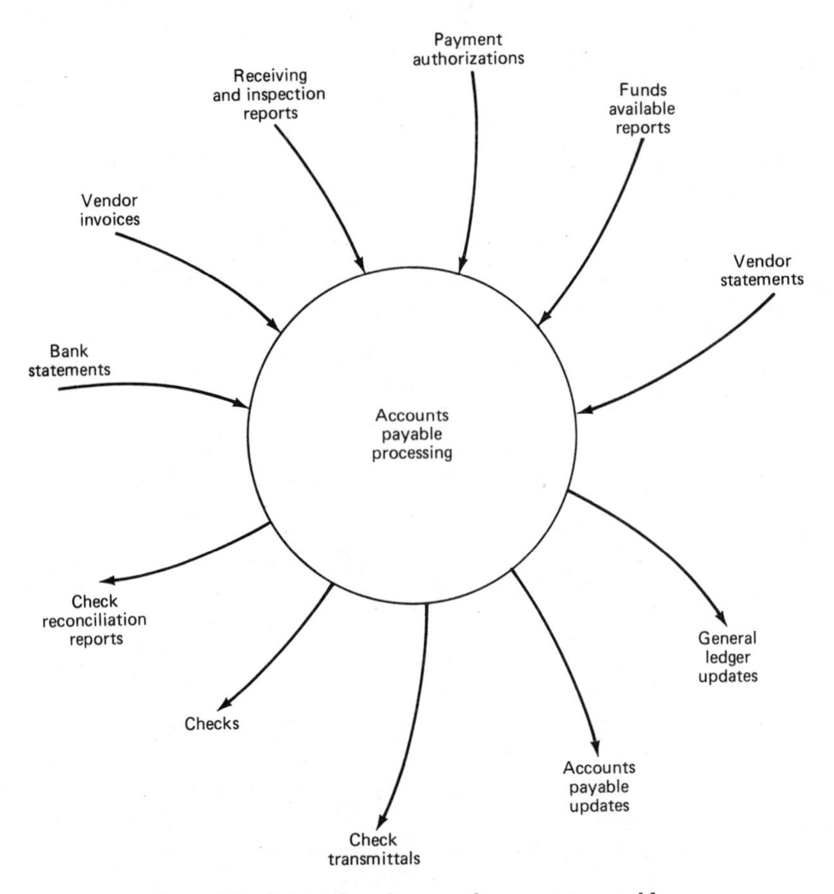

Figure 19.12 Context level data flow diagram for accounts payable.

types of terms and conditions which govern the invoices which the company issues for its products apply to the invoices it receives.

Payables processing normally requires that the invoices be segregated into those which offer discounts and those which do not, and those which have been authorized for payment (the goods or services delivered and accepted) and those that have not yet been authorized (because the goods or services have not yet been delivered or accepted, or because of some dispute with the vendor).

Based upon a variety of factors, including availability of funds, the firm will prioritize the invoices and issue payments for them (either in full or in part). These payments may be in the form of checks or some other funds transfer mechanism.

For each check that is drafted, a record is kept which contains its number, amount, date, payee, and account codes. Each check issued in this manner is printed on a report or kept in automated files. When

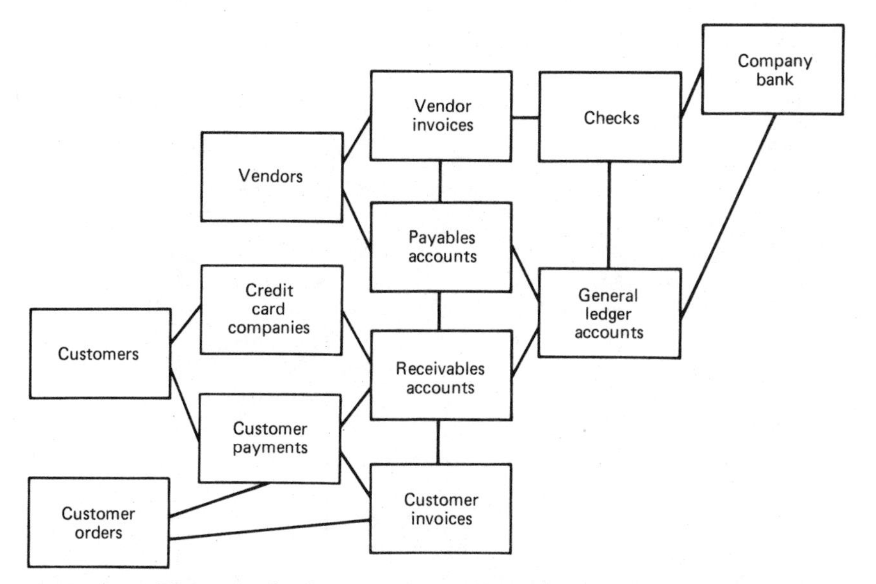

Figure 19.13 Enterprise level entity-relationship model—accounts receivable and accounts payable.

the bank statements and the corresponding canceled checks are received, they are reconciled with the records which the firm kept of its issued checks.

As with the receivables side, the firm normally keeps a ledger account, one for each vendor, which details all invoices received, payments made to the vendor, any discounts taken, and codes or other entries which indicate which area or operating account of the firm is to be charged for the amount of the payment involved.

General Ledger (Figures 19.14 and 19.15)

Each firm must keep a detailed set of financial records in which all financial transactions and any changes to its assets and liabilities are recorded. These records are organized into detail accounts, which are numbered in such a way that a hierarchy of accounts is formed. Those accounts which are higher in the hierarchy represent summary balances for the more detailed accounts below them.

These accounts are normally set up and organized by the firm's internal accounting personnel or by its external auditing firm. Each account represents the detailed transactions for some aspect of the firm's operation, such as office supplies, payroll, taxes owed, materials purchased, goods sold, etc.

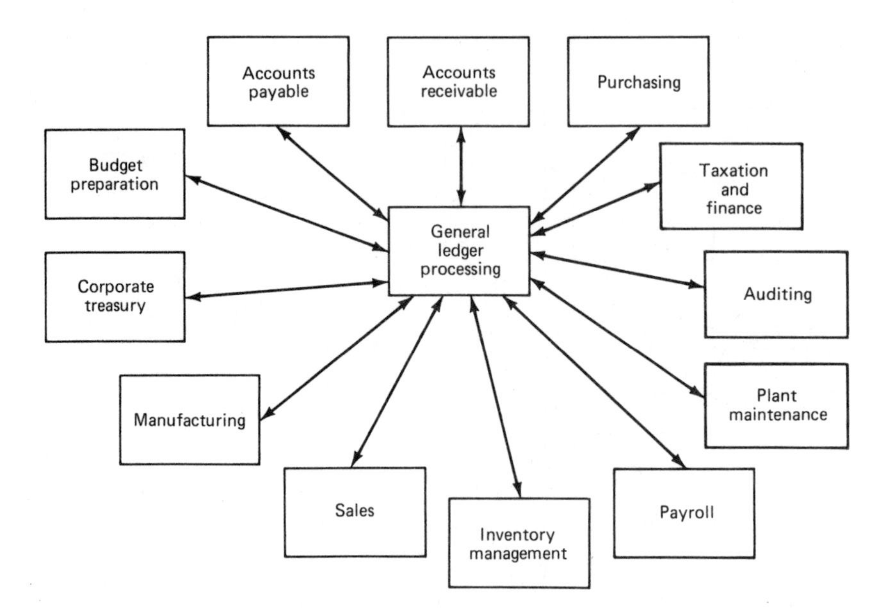

Figure 19.14 Process flow for general ledger.

There are normally detail accounts for every type of asset and liability of the firm. Most firms use a double-entry method of bookkeeping, where each entry is recorded twice, once as a debit (amount owed, subtracted, or removed) and once as an offsetting credit (amount received or added).

Thus, if a payment is made to a vendor, a credit entry will be made to the vendor's account record and a debit entry made to the cash account. Still other entries will be made to the various internal accounts which will be "charged" for the payment against their operating budgets.

Most firms develop one or more detail and summary budgets which are used to monitor income and expenditures. These budgets are prepared for a period ranging from a week to a year or more. They may be organizational unit budgets for administrative purposes, project-specific budgets for development projects, manufacturing budgets, sales budgets, etc.

Each budget is named and listed to the level of detail desired for each expense associated with the budget and each item of income expected to be received. As the time period or project schedule progresses, the actual income and expenditures are recorded and any variances (differences) between actual income and expense and project income or expense are noted and explained. From time to time, budgets may also be revised based on new projections.

Figure 19.15 Context level data flow diagram for general ledger.

Budgets are used primarily for management and control purposes. Many budgets are developed from predetermined numbers (funds allocated to a specific project), and the budget shows how those funds are to be spent and how they are actually being spent. The firm can use the variances to determine whether to allocate more funds or to cut expenses, as the situation dictates.

Marketing and Sales (Figures 19.16 and 19.17)

Most activities of the firm will be initiated from the marketing and sales units, which operate in distinctly different ways

A marketing unit seeks (a) to determine customer needs and to develop products that satisfy those needs, (b) to identify and develop new markets or new uses for company products, (c) to develop ways to sell

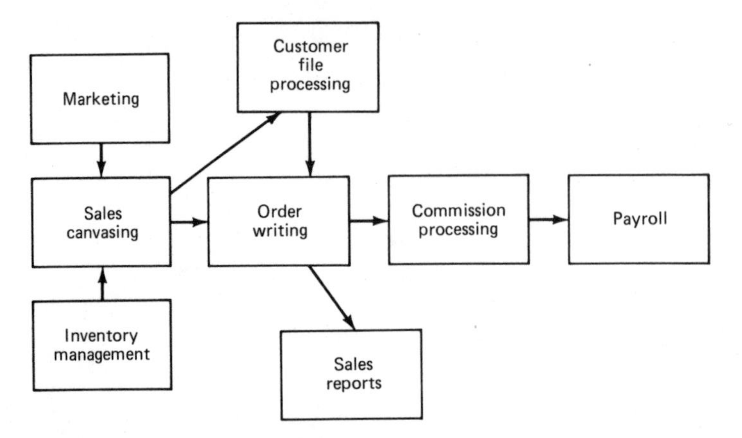

Figure 19.16 Sales processing flow.

products in old markets, or (d) to determine how existing products need to be modified to meet changing customer needs.

A sales unit seeks to sell products to customers and generate income for the firm. The sales unit is normally the first to identify and contact specific customers and is responsible for introducing the product to the customer, inducing the customer to buy or carry the firm's products; to take orders from the customer; and to try to induce the customer to carry larger quantities of product. The sales unit is the primary customer contact point, and it is the salespeople who normally are charged with making the first try at resolving customer problems or complaints.

Marketing units conduct surveys of potential customers, and develop and test market new products. Much of the work in the market-

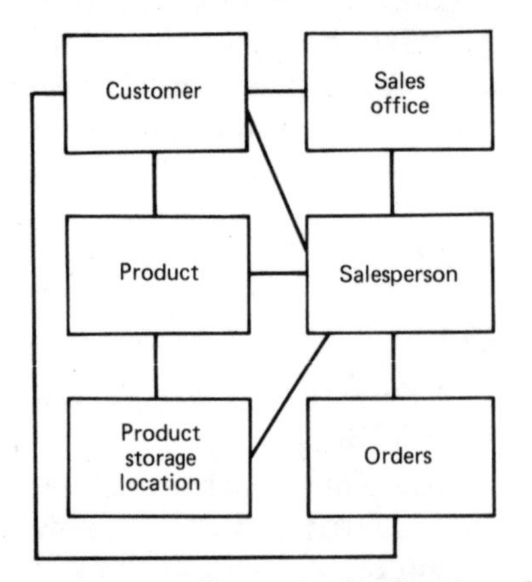

Figure 19.17 Enterprise level entity-relationship model for sales.

ing area is statistical in nature and is used to develop customer profiles, sales and economic trends, industry trends, etc. Marketing will normally be charged with determining the sales price of any product introduced and the way in which it is to be presented to the customer. Marketing in many cases determines the level of quality to be built into the product.

Sales units are concerned with actual sales, developing sales projections and estimates, and monitoring actual sales against those projections. It is normally the sales unit which provides sales figures to manufacturing, which in turn adjusts production and purchasing schedules.

Sales also presents to management its estimates of the level of revenues which can reasonably be expected; these estimates in turn are used by management to determine the amount of money it will have to spend or borrow to fund operations.

The success of any firm is highly dependent on the success of its marketing and sales organizations.

As orders are generated, commission amounts are determined and allocated to the sales staff. These figures are then routed to payroll for compensation purposes. In applicable cases, sales reports also provide the basic information for any licensing or royalty payments which may be due on the products sold.

Sales estimates, in terms of units and dollars or other measuring criteria, are compared to the actual sales performance in a manner very similar to the financial budgeting process. Many different reports are generated which compare sales on a period-to-period basis, on a year-to-date basis, by product, by sales area, by salesperson, by product line, against competition, etc.

Payroll and Personnel (Figures 19.18 to 19.20)

All companies have employees, whether one or hundreds of thousands. The firm maintains records on these employees from initial employment application to termination, retirement, death, and sometimes even beyond that. These records may span 1 or 2 days, or decades and contain all employee information which the firm feels is relevant to its business. This information may include

- Previous job history
- Personal history
- Medical history
- Position, transfer, promotion, and performance history within the company

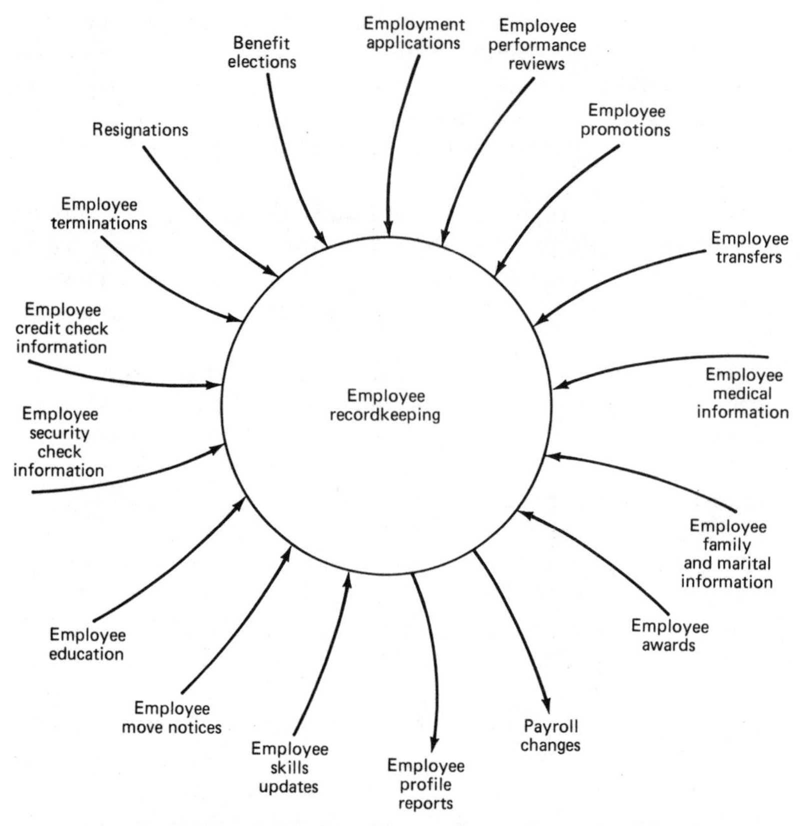

Figure 19.18 Context level data flow diagram for employee record keeping.

- Salary and bonus history
- Educational history
- Skills possessed by the employee, and the level of those skills
- Information about the employee's family
- Any benefits the employee may elect
- Employee residence information
- Employee taxation information
- Credit and financial information
- Arrest and conviction information
- Union affiliation
- Hobbies and outside interests
- Investment information

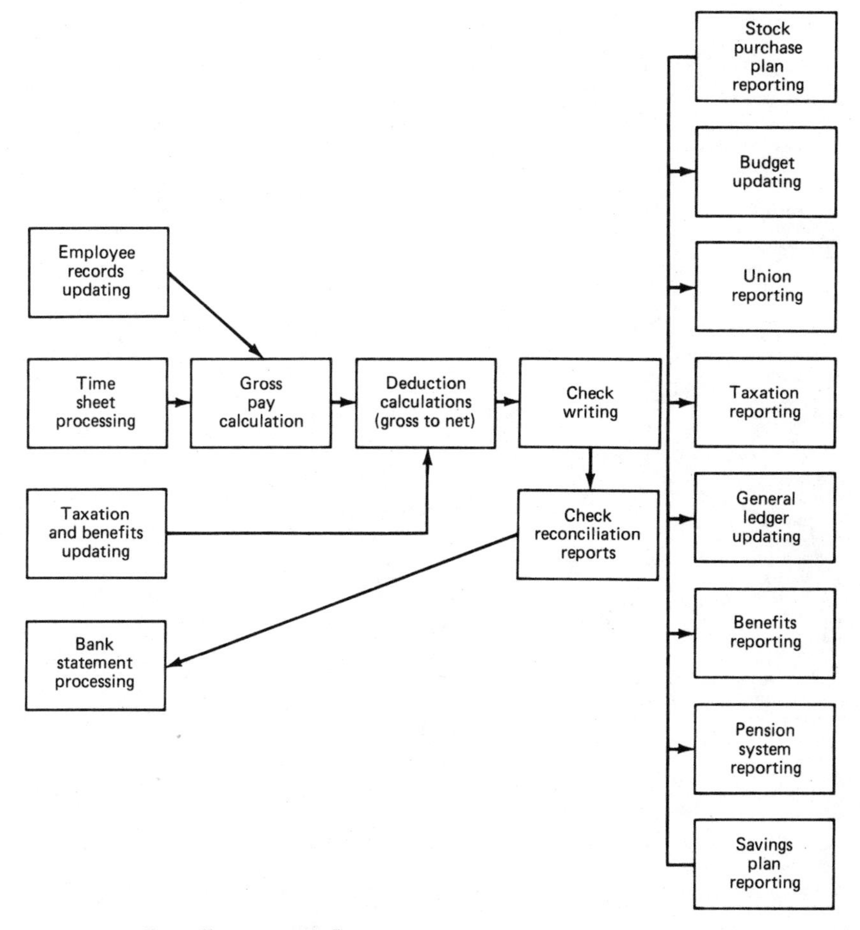

Figure 19.19 Payroll processing flow.

- Retirement information
- Participation in outside organizations or other companies

Although the firm may acquire any information about the employee which is needed for employment purposes or for general background information, there are very strict regulations as to what it may do with that information, and how and when it may disclose it. Aside from product information, company strategic planning, and executive-decision making information, employee information is the most sensitive and by far the most strictly regulated. If the firm has any data security procedures in place, they are likely to be in the human resources area. The firm is more likely to be open to legal proceedings because of its personnel practices than from any other activity in

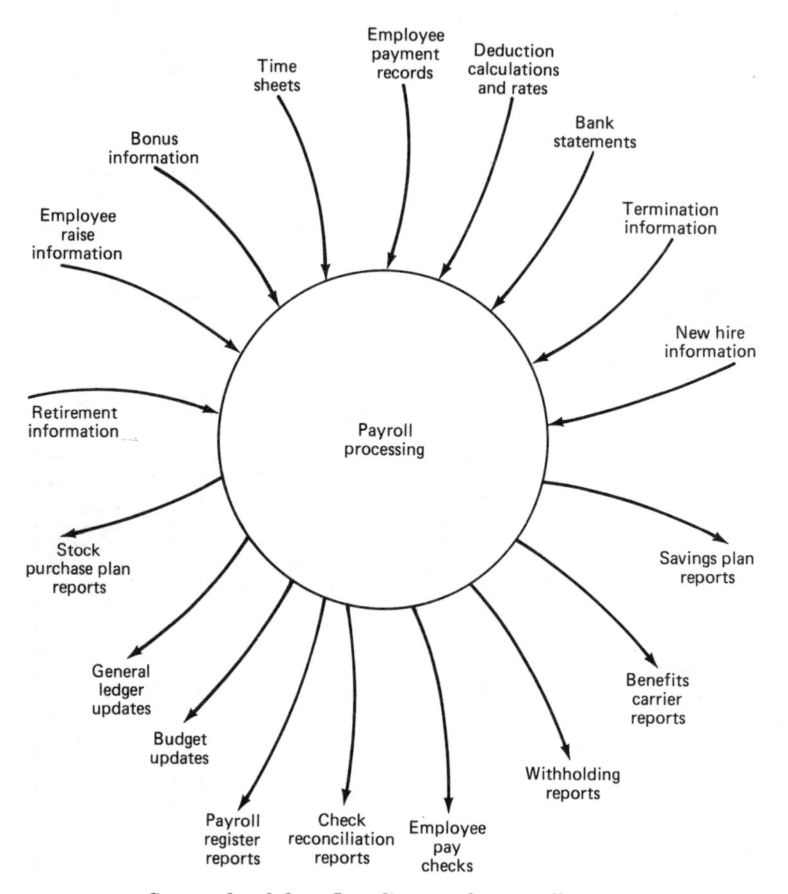

Figure 19.20 Context level data flow diagram for payroll processing.

which it engages. Many firms have more systems in place for human resources than for any other unit, and these systems tend to be among the oldest in the firm's systems portfolio.

The payroll and compensation systems receive input from time sheets, commission reports, bonus reports, etc., and generate paychecks. All moneys due the employee are totaled, and a gross pay amount is generated. All tax and other deductions are computed, and a net amount is determined which is then paid to the employee.

Depending upon the number and variety of payment methods, the number and variety of the deductions, the number of locations where the employees work, and the taxing authorities for which money must be withheld, these payroll computation programs can be among the most complex the firm has. They are also among the most unstable, in that benefits and other deduction information, as well as taxation information and rules, change very frequently.

As the payroll checks are being cut, reports related to budgeting, taxes withheld, and benefits and insurance provided are generated and sent to many areas of the firm, e. g., accounts payable, finance, and general ledger. The moneys withheld for these deductions are placed in accounts for payment to the government and for payment of various other benefits. Moneys are set aside during each payroll cycle for pension purposes, for employee stock purchase plans, for employee savings plans, etc.

As with accounts payable, check registers are printed for later use in check reconciliation systems. All employee detail deduction information and year-to-date information is maintained for end-of-year tax reporting and for year-end statements. The information is also sent to finance for inclusion in the firm's quarterly and annual financial statements.

Chapter

20

Case Study Illustrations

CHAPTER SYNOPSIS

The case studies in this chapter reflect the environments and problems which face real companies. Each case either is loosely based upon a real company, is an amalgam of several companies of the same type, or illustrates certain characteristics of that type of company. Each case reflects a different type of organizational environment. The case material should be used in conjunction with the chapter on the major types of applications so that the analyst can compare the application as presented in the company with the general application description.

The cases depict the applications and the various types of interrelationships between those applications in each business environment, the common data entities, data entity relationships, and data entity attributes. They also illustrate the most common processing sequences and functional areas involved.

Cases

- Apex Biscuit Company
- National Association for the Advancement of Data Processing
- System for Centralized Library Order Processing Services
- Studentext Publishing Company
- Last National Bank and Trust Company
- Verylarge Full Service Brokerage Company

Notes and Guidelines on Case Study Use

These case study materials provide selected background information on the representative firms in the basic types of industries in which an analyst can be expected to work. The information presented is similar to information which would be given to the analyst at the start of a new assignment. There is no implication that the information is either correct or complete, simply that it corresponds to the understanding of the analyst's manager who is providing the briefing.

Although each case is based upon a real company and a real systems problem, the information presented represents only a part of the overall company structure, operations, and functional organization. This corresponds to the application orientation of most projects. Case studies are a training and teaching aid, and a vehicle for understanding company complexity and conveying the necessary information. They are not meant to necessarily represent reality, but to provide the analyst with a "flavor" of the business. Cases have been structured such that the company activities or operations under discussion should fall in the realm of the reader's general knowledge.

Our discussion of common functional applications in the preceding chapter dealt with managerial and administrative functions. Those functions, although implemented differently in each type of organization, have common threads. Whereas the application discussions focused on those basic aspects of the functions which are similar between companies, the following cases focus on company-to-company differences.

APEX BISCUIT COMPANY

Background

The Apex Biscuit Company is an old and established firm which makes a wide variety of packaged baked goods. The firm employs approximately 5000 people nationwide. The firm is headquartered in the same large metropolitan area where it was originally founded in 1892. Its products are manufactured in six baking plants strategically located throughout the continental United States. The six bakeries supply a total of 150 firm-owned warehouses, each of which is located near the center of the sales territory it serves and which facilitates its distribution networks. In addition, each bakery leases space in public warehouses which act as auxiliary storage areas.

The Bakery and Warehouses

The company has divided the country into six regions, each serviced by a bakery. Each region is further divided into sales and distribution territories, each serviced by a centrally located warehouse. This structure allows the firm to service the 48 contiguous states plus parts of Canada and Mexico.

Although the firm is national in scope, its sales, and thus the locations of its bakeries and warehouses, reflect its growth and expansion history, with four of its six bakeries located east of the Mississippi. These four bakeries also service 95 of its 150 warehouses. Combined, these bakeries and their associated sales regions account for more than 80 percent of the company's gross sales.

Each bakery produces its own subset of the complete line of products, which are locally baked, packaged, and shipped to its warehouses using a combination of public freight companies, leased trucks, and a fleet of firm-owned trucks. The specific subset of products and packagings is determined by the local marketing and sales organization; quantities are determined by orders processed by that bakery.

Each warehouse has a fleet of delivery trucks whose drivers also act as order takers from established outlets on their delivery routes. The warehouses also act as the base of operations for the firm's sales force whose job it is to secure new customers and who are paid a commission for the initial sales and an ongoing sliding-scale commission based on all orders from customers they have signed up.

All sales within the area serviced by the warehouse and its sales force are the responsibility of the warehouse managers. Once a day, each warehouse manager consolidates all orders from his or her sales force and places a bulk order for products with the supplier bakery. These orders determine what products are produced by the bakery, in what quantities, in what packagings, and in what mix.

In some instances, a bakery may also act as a warehouse and ship directly to the retail outlets using trucks based at the bakery. In these cases, the sales are attributed to the bakery itself.

The Products

Apex Biscuit produces approximately 175 different product items, ranging from saltine crackers through tins of specially produced cookies which are sold only during the Christmas and Easter seasons. Each product item may be packaged in a variety of ways. For instance, saltines, one of the firm's staple items, may be packed four to a glassine package and sold to restaurants, diners, and other institutions, and may also be packed in 50-pound bags and sold (predominantly in the south) for sale from old fashioned cracker barrels. Between the 4-pack and the 50-pound bags there several dozen sizes of packages, boxes, bags, cases, etc.

Since the shelf life of its products is limited and its profit margins are low, the firm's overall profitability is determined by how closely its production matches its sales and by the amount of over-age product which must be removed from the retail shelves and written off. Those products which are removed from the shelves are either sold at company owned thrift stores at deep discount or given to various charitable organizations which feed the needy and homeless.

Although the firm is managed from its national headquarters, in effect each bakery and the warehouse network it is responsibile for acts as a semiautonomous unit determining what it will produce and sell. Production figures—by product item, sales, returns and removals, and production and sales expenses—are reported on a monthly basis to the home office for roll up to the corporate level.

Sales Issues

The firm's profits, while acceptable, have been relatively stable for the past few years and have shown little sign of growth. The firm has decided to embark upon a program which is aimed at increasing both its sales and its profits. As its first step, it has hired a national sales director whose first task will be to determine where the problems lie, if any, and where the firm should direct its efforts toward achieving its new profit goals.

There are a number of areas which appear to be fruitful areas for examination, each of which offers a number of alternatives.

At the warehouse level, the firm can (a) reduce the number of warehouses and consolidate them, (b) relocate some or all of the existing warehouses, or (c) open new warehouses.

At the product level, the firm can (a) reduce the number of products being produced, (b) reduce the number of different packagings, or (c) reexamine the product pricing structure.

At the sales force level, the firm can (a) make the salespeople responsible for ongoing order taking as well as initial sales, (b) eliminate the sales force and allow the delivery truck driver to make initial sales, (c) reduce the number of salespeople, (d) restructure the sale commission schedules, or (e) install a sales quota system to promote sales.

At the bakery level, the firm can (a) assign specific products to specific bakeries, (b) remove responsibility for the warehouses from the bakeries and establish separate sales regions, (c) eliminate some bakeries altogether, or (d) relocate some bakeries and consolidate regions into the remaining bakeries.

The new sales director, after some initial examination, realizes that important information about the company's sales performance is missing. After doing some initial data gathering, the staff determined that no consistent information exists on all the bakeries. An initial analysis uncovered the following additional facts.

1. Although each bakery knows what products it produces and how those products are packaged, there is no firmwide consolidated list of all possible products and packagings.
2. There seems to be no consistency in the reporting of the size or volume of warehouse sales or the size of the territory served by each individual warehouse. In addition, since, for economy, payroll and commission checks are prepared at the individual bakeries, there is no consolidated list at headquarters of salespeople or their individual performances. This lack of consolidated data makes bakery-to-bakery, warehouse-to-warehouse, and salesperson-to-salesperson comparisons almost impossible.
3. Bakeries report production on an aggregate basis. That is, all sales are reported as net dollars. There are no individual figures for units produced, returns, etc. Because of the high product turnover and short shelf life, no ongoing inventory is kept at either the local warehouse or bakery level.
4. Sales commissions are paid on the basis of ordered quantities, and no adjustment is made for returns.

Because populations in the sales territories vary, there is no consistent way to measure relative sales performance. The variations in product designation and ordering methods make order analysis impossible. This difficulty is compounded by the very heavy volume of orders and the large number of verbal orders which are accepted and filled from the delivery trucks. The trucks themselves are loaded with a variety of products based upon historical demand, and the trucks are audited on a net basis.

The sales force commissions are computed each month, and although all salespeople are paid on a commission, the commission rates vary from bakery to bakery and from territory to territory. This is mainly a result of the size and profitability differences between territories.

Production Issues

Because their products have a short shelf life and thus need to be baked fresh each day, each bakery must order its raw materials from local sources. The wide variety of products and the use of locally available materials has led to wide variations in product ingredients. Although these variations are most apparent between plants, they can also be seen within each plant as sources change, and are in some cases based upon seasonal variations in the types and availability of the raw materials. There are also location-to-location variations in package printing and in the product's taste, which is due to differences in the flours, flavorings, and oils and shortenings used.

The differences in age of the various plants has led to wide variations in the levels of plant automation and in the age and types of machinery used in the manufacturing process. There are also distinct differences in information availability in each plant. Some of the newer plants have relatively modern data processing facilities, and correspondingly modern information systems. Others operate with outdated or even obsolete equipment.

The traffic control systems in each plant are also very different. In some plants and warehouses the trucks are routed so that each truck can visit as many as 30 points a day, while others can only visit as few as 5 or 10. Inventory management of raw materials in some plants allows for minimal inventory levels and efficient operations, while others have large overstocks of some materials, while shortages in others delay bakery production of some items.

Because of the heavy volume of deliveries and returns, and the corresponding heavy volume of orders, both written and verbal, each warehouse and bakery maintains its accounts for its distribution points on an open account basis. All goods are delivered to and sold by the retail outlet on a consignment basis. Order transactions are added to the open account, and returns are subtracted on a daily basis. The retailer is given a weekly statement of deliveries and returns on a product-by-product basis, as well as a bill for net moneys due from sales. All discrepancies between retailer records and company delivery and return records are settled on a net basis without regard to individual products sold.

Since the company advertises heavily on both local and national media and since it makes heavy use of discount and cents-off coupons as a sales promotion mechanism, the retailer may pay part of its bill for products sold in the form of redeemed coupons. Since coupons are on a product- and sometimes on a package-specific basis, the retailer must accompany any such coupon payments with specific delivery receipts for the coupon products and must verify that it has sold coupon products in the amount being redeemed. For each coupon redeemed, the retailer is paid 5 cents over the coupon face value amount. The retailer is responsible for ensuring that it does not accept coupons which have expired.

The retailer may sometimes run specials of its own, in which case, it will either bear the entire cost of the difference between retail price and the price it sells the products for, or in some cases it will share the cost with the company. This sharing may reflect just the cents-off cost, or may reflect the cost of advertising as well.

Because of the heavier demand for product which these sales engender, the retailer usually coordinates these sales with the company to ensure that production and delivery runs are adjusted to ensure proper stocking levels. In these cases, the company does not accept returns of product, and any expired stock is absorbed by the retailer.

NATIONAL ASSOCIATION FOR THE ADVANCEMENT OF DATA PROCESSING

Background

The National Association for the Advancement of Data Processing (NAADP) is the oldest and most prestigious organization of data processing professionals in the world. It was founded in the United States almost 40 years ago; today it has members and branch organizations in 50 countries around the world and is affiliated with similar national organizations in 40 more. The organization is a not-for-profit organization, chartered to advance general understanding of data processing, to promote the interests of the data processing community, and to provide a forum for the discussion of topics of interest to the data processing profession.

As with most professional organizations of its kind, it is managed by officers and committee heads elected or appointed from its general voting membership. There is a national headquarters which is staffed by a small group of paid professionals, whose job is to maintain records for the association, process membership renewals, collect and bank all dues and fees, publish its various periodicals, arrange for its semiannual trade shows, and answer questions from members and prospective members.

Internal Association Organization

Although it is a national organization, the basic functions of the association are performed by a complex of semiautonomous local chapters, which are located in most major cities around the United States. In addition to the local chapters there are student chapters attached to most major colleges and universities.

Each local organization is headed by elected and appointed officers (mirroring the national organization structure) who are elected by and from local membership and who work on a volunteer basis. These local organizations take general direction from the national governing body and receive the bulk of their funding from a redirection of a portion of the national dues. In addition, the local organizations may supplement this funding with local dues collection and fees and with revenues from local activities.

In addition to the formal chapter organization, the association promotes and supports some 60 national-level special interest groups (SIG), each of which is devoted to a specific aspect of data processing theory, technology, or management. The local chapters may also optionally sponsor local SIGs which may be similar to their national counterparts or which may reflect purely local interests. Funding for national SIGs is obtained from a separate dues structure, which is collected at membership renewal time by the national organization. Local SIGs are responsible for raising their own funds, or they may be funded in whole or in part by the local chapter.

The NAADP maintains reciprocal agreements with other data-processing–related national organizations. Under these agreements, publications, functions, and services of the national organization, and membership in the local chapters and the national SIGs are open to members of these other organizations at NAADP member rates.

The publications, functions, and services of the national organization, and membership in the local chapters and the national SIGs are also available to non-NAADP members at higher rates than those offered to either NAADP members or to members of affiliated organizations.

Publications

The NAADP publishes a monthly membership magazine which is sent to each member as a membership benefit. In addition 10 monthly, bimonthly, and quarterly journals are also published and are available, at additional cost, on an annual subscription basis. Both the monthly membership magazine and the journals are available to nonmembers at a cost which is slightly higher than that charged to members. Back issues of all publications are stored at headquarters and may be purchased for a premium over the normal single issue cost. Reprints of selected papers and articles may also be purchased, either individually, in bulk, or bound by topic.

The headquarters also arranges for the publication and distribution of the magazines of the national SIGs. Publication of local chapter and local SIG publications are handled by the local organizations.

Mailings

The headquarters assists the chapters and SIGs by supplying them with various membership listings and by printing mailing labels for their publication and advertising needs. As the keeper of the national mailing lists, the association also raises funds by selling copies of the list, in label form, to organizations connected with data processing and approved by the elected board of directors. These lists may be segregated geographically, by interest type, or by any other available selection criteria. The more selective and specialized the list, the higher the charge.

The Membership

The association currently has 100,000 members nationwide, of which 25,000 are student members. For historical as well as practical reasons, the association's regular membership renews on a calendar year-end basis, while the student members renew in September. The regular membership nonrenewal rate is 20 percent, while the rate of new member pickup is around 22 percent. The student member renewal rate reflects the normal 4-year student status. Thus each year, there is a surge of new student members (about 25 percent of the total), while a corresponding 25 percent leave student status. Of those leaving student status, approximately 50 percent convert to regular membership while the remainder drop membership entirely.

Current Processing Environment

Despite its data processing orientation, the availability of membership expertise, and its size, the association, like the shoemaker's children, lacks all but the most rudimentary processing capability. A large portion of its membership records are kept in manually maintained, paper files. A number of years previously, in an act of desperation, the organization's mailing list was given to a service bureau owned by one of its members, which developed a minimal system for mailings. This system was developed gratis and is run for the NAADP at cost.

The bureau-developed system transferred the master file to tape where it is processed in a batch environment. Each record contains the member's name, address (two lines), codes which reflect membership renewal date, and codes reflecting the number of copies of each publication ordered.

Each record is identified by a unique membership number. Renewals and new memberships are punched onto cards which are then used to update the master file.

The update system produces a combined proof list containing all new members, all deletions, and all renewals. Every quarter a master list of all members with their renewal dates is compiled for headquarters staff.

The NAADP governing council, recognizing that the effective limits have been reached for both the existing manual and automated systems and being desirous of expanding both the membership and the number and level of benefits and services available to its members, has commissioned the newly hired headquarters data processing manager to perform the initial analysis and develop a design for a more up-to-date automated system which would bring the association's processing back in-house and which would provide flexibility and expansion capability for the association through the rest of the century.

Issues

Because of the nature of its membership and because of its largely decentralized operations, many of the local groups have developed their own systems for record keeping purposes. These systems are maintained on the machines available to the members, which range from mainframes to personal computers and cover the range of hardware vendors in all categories.

Although the national office maintains its records on an up-to-date basis, the local systems vary in how current they are. Many systems use nationally produced lists to update their own systems' files.

Because local groups can accept members who are not members of the national organization and because they also maintain people on their lists who are not members but who are of interest to the local group, there is a great discrepancy between the national and local versions of these various lists.

The batch nature of the national system and its lack of completeness and flexibility greatly hinders the national organization's ability to provide current information in response to member queries. Failure of a member to renew causes the record to be dropped after two months. Since many members renew late or skip a year entirely between renewals, each late renewal for whatever reason requires the entire membership record to be reentered. This process of dropping and reentering causes much valuable historical information to be lost between cycles.

The batch nature of the system and the delays inherent in the processing cycles, along with the long lead times in publication preparation, cause many of the publications to have production overruns while others have shortages.

The association is required to provide an annual report to its membership on its financial condition and the financial condition of each of its related groups. These financial reports are chronically late, and in some cases, the national organization and its local units have come close to losing their not-for-profit tax-exempt status. In addition, since advertising and postal rates, the tax-exempt status of its publications, and the rates which it charges for its mailing list are all based upon independent audits of its circulation, and these reports are in turn based upon its mailing lists, the organization is continually lacking adequate backup when justifying its rates in each of these areas.

SYSTEM FOR CENTRALIZED LIBRARY ORDER PROCESSING SERVICES (SYCLOPS)

Background

In 1980, SYCLOPS was chartered by the state department of education to develop an integrated data processing system to automate the back office processes of its public library systems. SYCLOPS was to be governed by a committee consisting of the directors of each of the state's 35 public library systems. Library back office processing covered all activities from reviewing initial book order lists from the publishers to the final payment to the publishers. All of the intermediate processing was to be automated to the extent possible, and the state's intent was that SYCLOPS take over as much of this work as possible from the systems and the individual libraries, thus freeing the staff for more productive work.

An additional goal was to simplify the processing and achieve any economies of scale possible in terms of bulk ordering of books and materials, and in terms of processing labor costs. An additional goal was to provide for greater levels of standardization between the libraries and systems and for new and improved services which result from the availability of improved and centralized information.

Library back office processing includes all the work necessary to place an item on the library shelf for circulation. The process begins with advance publication lists being received from the publishers, by receipt of issues of forthcoming books, or by a request from a library patron for an item which is not currently in the library's inventory but which is still available from some source.

Lists of new items are previewed, review copies are obtained, and read by the librarians; a determination is then made of how many of each item is to be ordered. Orders are placed with the publication source or with publication wholesalers or distributors. When the items are received, they normally arrive with special library bindings; if not, they are sent to binderies for suitable rebinding. Once received, or rereceived, at the library, the books are reviewed by the cataloging staff, and the information necessary to produce the set of catalog cards is obtained. The card sets are manually typed, along with card pockets and spine labels, which are affixed to the item, and a protective cover is then applied.

In addition to the physical processing of the item, the back office also processes vendor accounts payables and inventory. This back office processing may be performed either in the individual branch, if it is large enough to warrant the staff, or at the system level. A library system usually includes all branches in a county, although depending upon the population, a system may encompass several counties, or a county may have more than one system. The more normal case is for the processing to be handled on a system basis since advantage can be taken of economies of scale. If performed at the system level, it may be in the main branch or in some independently located office devoted to this processing.

SYCLOPS was intended to extend this concept one step further, using its system to perform this processing on a statewide basis, gaining further economic leverage with the suppliers and further economies in the processing itself.

The Products

Public libraries stock a wide variety of items, the predominant item being books, both fiction and nonfiction, and reference volumes. In addition, the libraries stock periodicals; records, tapes, and compact disks; video tapes, movies, and educational films; presentation materials; artwork, etc. Regardless of type, all items are cataloged, which involves reading or otherwise reviewing the item and generating a brief synopsis, usually consisting of one or two lines, of the contents or an equally brief description. A set of catalog cards is generated with item descriptive information. Each set contains a card to be filed under the title, the author's last name, and each of the selected subject categories. Fiction items are usually shelved by author's last name, while nonfiction is shelved by subject. Nonbook items are shelved by title, topic, or category.

There are a number of systems for identifying library items. The primary, and most universal, one is known as International Standard Book Number (ISBN) and involves a multipart format, 10 digits in length, which identifies language, publisher, book identifier within publisher, and a check digit. The ISBN is unique for a particular title, edition, and binding. The ISBN is assigned by the publisher.

A second numbering scheme is known as the Library of Congress number, which is assigned by the Library of Congress cataloging staff in Washington, D.C.; each Library of Congress number is unique with respect to title and edition, but not necessarily to binding.

A third numbering scheme is known as the Dewey Decimal System; this number is assigned by the library cataloging staff after determination of the volume's primary subject matter. The Dewey number is the number by which nonfiction is shelved. The Dewey number ranges in value from 001 to 999 and may be followed by any number of decimal digits which further identify the book's subject matter. Most American volumes carry all three numbers.

Because many library items are not books, many libraries have developed item identification schemes to identify, catalog, and control these items. There is little if any standardization of these other schemes among libraries or systems.

Library Accounting

Public libraries are funded by federal, state, and local taxes. To make most effective use of these funds and to avoid duplication of effort, the libraries of the state are aggregated into systems. Each system is composed of and controls the branches located within its library district. Each system is autonomous and is responsible to local legislatures. Each library receives funding for operations and new acquisitions from the local and state legislatures. Funding is also generated from overdue item fines, from donations, and from membership dues charged to patrons from outside the branch district who wish to make use of the library's facilities.

Each library must prepare and present to the system a yearly budget of operations. This budget must include all fixed and recurring expenses, such as salary, rent, lease, or mortgage payments, utility costs, building maintenance, and new item acquisition plans.

The income side of the budget usually includes known donation funds, an estimate of membership dues, and estimates of state and local funding. The state and local funding is based upon a formula which factors in library district population and circulation numbers. These individual library budgets are rolled up to the system level, and the final numbers are presented to the state and local legislatures for approval. Once approved the numbers are included in the governmental budgets as a separate assessment for libraries.

Once the funds are approved, the library may draw on them for its operational and acquisition expenses. As items are purchased for its collections, the invoices are vouchered and passed along to the system for payment. The systems in turn maintain ongoing acquisitions accounts for the libraries and make up monthly and quarterly account status summaries.

Each library, or system, will make up its own purchase orders for items for its collections and include its budgetary line numbers as reference. As the vendor invoices are submitted, the library will ensure that those budgetary line numbers are included when they are vouchered.

Interlibrary Loan Functions

Items are freely exchanged between branches of a system, and a patron can withdraw or request any item within any branch of his or her home system. In addition most of the systems subscribe to and maintain an intersystem loan capability which in effect makes available to any patron all items in every branch in the state.

Currently all back office, inventory, and intra- and intersystem loan processing is handled at the system or branch level. Some systems and some of the larger branches have achieved some level of automation; however, there is no consistency among these automated systems. The commissioner of libraries for the state, an employee of the state education department, has persuaded the state legislature to take advantage of federal funds available for the purpose and to provide supplementary funding to establish and support the SYCLOPS organization, whose primary mission would be to perform in-depth analysis of the library systems' processing and to develop requirements for the design of an implementable automated system.

STUDENTEXT PUBLISHING COMPANY

Background

The Studentext Publishing Company is almost 100 years old. Its main product lines focus on textbooks and other educational materials, although it has a rather substantial popular fiction list.

The firm has grown over the years by acquiring other smaller publishing firms with similar or complementary lines, and today publishes under almost two dozen mastheads. Its offices are located in New York City; however, its warehouses and processing plants are located in Pennsylvania, Maryland, Massachusetts, Connecticut, California, Texas, Kentucky, Ohio, Tennessee, and southern New York State. Its data processing center is in the Tennessee plant with leased lines to the New York Development Center. Its customer service lines are connected to each plant and warehouse.

Studentext's first major automation was completed almost 10 years ago. At that time it was considered a model of its kind and was used as a showcase installation by the hardware vendor. Although it is still functioning adequately, it does not satisfy all of the firm's current needs and is not amenable to modification.

After much deliberation the firm has decided to reexamine its requirements and develop a new automated system, to move it from an essentially third generation environment to a more technologically up-to-date online database environment.

At the same time the firm wishes to add some of the functionality which it feels is missing from the present system and to provide it with more flexibility in the sales and marketing areas.

The Business

Although identified as a publishing firm, Studentext does not actually manufacture its own products. That is, it does not maintain any printing or binding facilities. Management categorizes the firm as a marketing and distribution organization. It "farms out" the traditional production activities and functions to contractors. Its own activities are centered on the front- and back-end work.

Front-end work consists of manuscript selection and editing, artwork processing, editorial services, and layout and typesetting. Once the final texts and galleys are approved, the work is sent out for production. Production consists of all printing, cutting, and binding activities normally associated with book publication.

Studentext acts as intermediary and contractor during the production process by arranging for completed flats to be picked up at the printer and transshipped to the company's bindery services. Once the finished product leaves the bindery, covers are affixed and the books are shipped to one of its warehouses where they are stored awaiting orders, or shipment to its outlet bookstores and other distributors. Other services performed internally consist of advertising, order processing, royalty accounting, billing, inventory, and other marketing services.

Product sales occur in a number of ways.

1. A commissioned sales force travels around to distributors soliciting orders for forthcoming titles and reorders for already released titles. These distributors sell on a consignment basis.
2. Other sales specialists travel to schools and universities to show school administrators and department heads texts and other educational materials in the hope that they will be adopted by the school or school system as standard text.
3. Some products are sold on a continuing subscription basis. They are serviced from the local plants, and product is shipped as it becomes available.
4. A new and growing market for Studentext is direct mail-order sales, which are initiated by advertising campaigns and by order cards inserted in various popular or specialty publications or by direct solicitation of the end customer.

The Products

Studentext has a large list of products for sale. In addition to the standard items such as fiction and nonfiction books, it has, as previously noted, textbooks and reference works. The following is a brief description of some of its different types of products.

1. Textbooks may be sold individually or in sets, as with encyclopedias. Other books may be sold in series; this is particularly true of texts where a school system may buy a reading series which consists of texts for K to 6, 7 to 8, or 9 to 12. Other series are sold to colleges, universities, and graduate and specialty schools.

In addition to the texts themselves, the series may also include workbooks, that are used and thrown away, textbook updates, and teachers' guides and texts. This class of product also includes worksheets and other visual aids such as maps and charts. In addition audiovisual aids may be sold as supplementary material.

2. Books may be sold to distributors, either individually or to chains of stores. The orders for these books may be placed (a) directly with the firm by either the branch or the chain main office, (b) through one of the firm's sales representatives, or (c) through some wholesale buyer who then resells to the retailer. Here the titles may be ordered individually, in bulk, or in what is termed a "dump," which consists of a collection of related titles, or similar fiction or nonfiction titles, sold as a unit along with the display rack and other promotional media.

3. The firm acts as producer and publisher for a wide variety of periodicals, professional journals, and other magazines. Most of them are sold either from newsstands or by subscription. In some cases, however, the firm produces the periodical for another organization, usually a trade or other specialty group, which provides it to their membership as a benefit or for a fee. Studentext maintains the subscription list for its own publications and provides subscription fulfillment services to its clients, on request and for a fee.

4. Certain lines of products are made available to the consumer through book clubs. The most popular of these work like the Book of the Month Club, where the consumer is enrolled and given a premium of a certain number of volumes at no or greatly reduced cost, and thereafter is offered a monthly catalog of discounted books from which the customer must select a predetermined number of items.

For each multiple of books purchased the consumer is offered credits toward more "free" books. Thus the consumer may get one book "free" for every five purchased at the "regular club price." The consumer must order a certain number of titles to cover the initial premiums.

The book clubs along with textbook sales account for a large percentage of the firm's repeat sales.

5. The final category of product consists of direct-mail sales. In this case the products are advertised in such a manner as to solicit sales directly from the consumer. The advertisements and their response cards are placed in various popular periodicals. The titles in this category are usually either specialty items, such as reference works, medical or other scientific works, or fiction titles. The products advertised in this manner are geared to the carrier medium, e.g., medical texts are carried in medical journals, etc.

Although this method of sales is more labor intensive and the initial sale is more expensive, it has the possibility of a large volume of repeat business at little, if any, additional cost. In addition, the firm can, through this method, build up an extensive mailing list of customer names and addresses along with certain demographic information, such as title and subject preference, which would cost a great deal of money if acquired from outside sources.

The Issues

The firm completed its initial automation project approximately 10 years ago. At that time the system was considered to be state of the art. It was an online system using traditional file access methods for data storage.

Much of the processing was performed in batch mode, and the applications which were developed at that time consisted of

1. Order entry
2. Accounts receivable
3. Back-order processing
4. Product master file maintenance
5. Inventory management

Author royalty accounting, because of its complexity, was and still is handled on a manual basis. The firm's list of titles and authors has grown substantially in recent years. Each book is contracted for with the author on an individual basis, and although certain similarities exist between contracts, no two contracts are the same. As a result of the individual nature and complexity of each author contract, and the subsequent difficulty in determining royalties due, the firm would like to automate this process and tie it into its sales analysis system, which is the primary source of information for royalty computation.

Since its initial automation, the firm has acquired many other publishing firms and product lines. The volume of sales and the number of products has far outstripped its own system's capacity. The firm's attempts to assimilate or adapt in some manner the systems of the acquired firms has so far met with failure. The firm has ordered a new generation of processor and has commissioned the development of a replacement system.

Management's directives state that they want a system which is (a) integrated, (b) flexible, and (c) capable of supplying the increased capacity required by the expanding business.

An additional objective, as stated by management, is to increase the firm's direct sales capability. Currently the firm must rent lists of prospective customers. These rentals are usually single-use only, and the prices for each list vary, depending upon the selectivity of the list.

The selectivity of a list is determined by the amount of demographic data known about each entry in the list. For instance, a list of all physicians in the country might rent for $.50 a name. The same list with type of practice indicated might rent for $1.00 per name. The same list with type of practice and specialty indicated might rent for $1.25 per name. If the firm were publishing a new text on cardiac surgery, it would obviously be more cost-effective to solicit orders from cardiac surgeons than, for instance, from general surgeons or pediatricians.

The firm has indicated that it would like to begin building a customer base containing the type of demographic information which would enable it to target its solicitations more cost-effectively and without the necessity of renting lists from outside vendors. The development of requirements for such a customer base and analysis as to how to best create one has been designated as a primary part of the first project.

Additionally, the firm would like to improve its customer service capability, its accounts receivable systems, and its inventory systems to reduce the number of back orders or nonfillable orders. In addition it would like to be able to have greater control over the distribution of product to its warehouse and storage locations, which will in turn reduce its distribution and delivery costs.

As part of this new system, the firm would like to automate and simplify its royalty accounting and payment systems. Currently each item must be accounted for separately, and the same author may receive payment for different items as much as 2 or 3 months apart. Those items with multiple authors, illustrators, editors, etc., receiving payment may be as much as 6 months in arrears. The firm must pay penalties for these late payments, and thus would like to eliminate the delays to the extent possible.

The company maintains a variety of customer and vendor accounting systems. Most of its sales outlets work from open accounts, with deliveries and returns being maintained on a continual basis and monthly accounting statements being generated. For chains, accounts are maintained in duplicate, with a master set being maintained for the chain home office and a duplicate set maintained for the branch location. Each branch is billed separately with consolidated statements sent to the home office.

Since accurate sales analysis is the basis for many of the firm's operational systems, including inventory management, production scheduling, royalty computation, sales commission computation, and distributor payments, an updated sales analysis system is also a high priority item.

Individual customers may open a variety of accounts ranging from cash prepayment through COD and credit accounts. Customers may also pay by credit card. For both individual and corporate accounts, the firm offers the option of master and standing orders and offers special shipment and billing options. All credit customers undergo periodic credit checks by the firm, and a customer may be offered credit facilities, or credit may be withdrawn based upon ordering and payment history. Because of the shift in emphasis to direct-mail marketing, customer accounting and billing is also a high priority item.

THE LAST NATIONAL BANK AND TRUST COMPANY

Background

The Last National Bank and Trust Company is a medium-size, regional, commercial bank with its roots in the New York metropolitan area. As with most institutions of its size it grew by acquiring other local banks and merging both their assets and customer bases into its own. Over time its assets have grown to about $5 billion, and its customer base has grown to approximately 500,000 account holders. Its branch network currently extends to some 500 locations, ranging in size from storefront offices to large modern complexes, many of which are in buildings designed, built, and owned by the bank itself.

The bank is considered rather conservative, waiting for its larger competitors to innovate new products and then modifying those products to its own needs. Its offerings fall into the traditional mold, with demand deposit accounts (checking), commercial and consumer loans, mortgages, and savings accounting for the bulk of its business.

The Last National has recently been bought by a large overseas bank, which resulted in a change in management and an influx of capital. The new management has set a strategic direction which is intended to streamline the operations areas, to provide a wider range of products and services offerings, and to improve the bank's image with its customers and with the general public.

As with most banks today, Last National has an extensive portfolio of existing automated systems. Many of them were developed as independent projects, and most are 10 to 15 years old. While the systems are still serviceable, the bank, to keep up with its competition, has embarked on a strategic change in direction with respect to systems development. The new management has funded a project to analyze the current account processing environment, with the ultimate goal of reautomating these systems in three areas.

1. From the existing multivendor hardware environment to newer, single vendor, more cost-effective machinery
2. From an account-specific processing environment to an integrated customer asset management environment
3. From a multiple file, batch processing environment to an online database-oriented environment

The bank would also like to take advantage of the efficiencies and economies afforded by personal computers, by reautomating, where possible, many of its smaller user-specific systems to these machines.

These development programs are dictated by the bank's need to provide more extensive and flexible financial services to its customers, by the expanding product offering environment, and by the concurrent need to assess the bank's exposure for a given customer across all its product offerings, and to manage its risk in a more efficient and economic manner.

The bank's competitive environment has also been changing over the years, and while it is still profitable, those profits have been relatively flat, even in the face of an expanding customer and customer account base.

In previous years the bank's competition has been other similar commercial institutions in the area. Many of the products offered were solely the province of commercial banks. With the changing regulatory environment, the bank is facing competition from many new sources. These include

1. Savings and loan institutions that now offer product lines very similar to those of commercial banks
2. The larger money center banks that are expanding their automated teller machine (ATM) and branch networks into what was Last National's territory
3. Other financial institutions, such as brokerage firms, that are

now offering products which compare rather favorably with those of the bank

4. Large retail chains which have acquired financial services units and are expanding their financial and credit services to the point where they now compete with commercial banks in many areas
5. Large insurance companies which have expanded in a manner similar to the large retail chains mentioned above
6. Foreign banks which have opened domestic branches in their own name and are establishing branch banking systems

The New Strategy

As the first phase of its planned reautomation project, the bank has decided to tackle four of its major systems requirements, two old and two new: installation of an ATM network, demand deposit, consumer savings, and a new central customer information system. These projects were chosen since they are tightly linked in terms of both processing and data requirements. Each project provides a component for implementing the bank's new strategic plan for coping with its new competition.

The ATM Network

The bank has recently installed an ATM network in its branch offices. These machines are available 24 hours a day, 7 days a week; they allow clients to make payments on loans and deposits to their various accounts; to obtain last transaction, interest, and current balance information; to transfer funds between accounts; and to withdraw funds from savings and demand deposit accounts.

These machines are activated by a credit-card–style customer identifier card with a magnetic stripe and a customer-entered personal identification number (PIN). Each of these machines has a small display screen, a telephone-style keypad, a cash dispenser slot, deposit slot, and a magnetic card reader slot.

The new machines have been installed so as to provide maximum availability to bank customers. In some locations the machines have been placed in branch lobbies and new card-activated access doors have been installed. In other branches, lobby machines have been supplemented by through-the-wall machines.

The machines themselves have been hooked to the central bank processors via redundant phone lines. In most cases each line to a branch has been routed through a different phone company switching office to provide maximum protection. Special procedures have been instituted to ensure round-the-clock servicing for the machines, including stocking counter deposit and withdrawal slots, and to ensure that the machines have sufficient cash and receipt tickets. Additionally the bank has set up a customer service department to staff the "hot" phones to each branch machine area.

Demand Deposit Accounts (DDA)

The traditional demand deposit processes were driven by a series of documents: the check itself, customer deposits, and special instructions from the customer (such as stop payment orders, etc.). Under the new environment these traditional items still exist; however, they have been augmented by new ATM transactions, such as direct cash withdrawal and payments, transfers to other accounts, and a host of new products. These products, usually associated with a traditional DDA account, include overdraft protection (a form of automatic consumer loan), money market accounts (for unused cash), debit cards, etc.

In order to become more competitive, the bank has also instituted such features as automatic bill payment, checkless checking, combined account statements, chargeless checking (in combination with minimum balances in other accounts of the same customer), and revolving credit accounts, which use the customer's home equity as collateral. These new services all utilize traditional bank services but combine them in new and innovative ways.

Consumer Savings

The traditional passbook savings accounts offered by the bank required that the customer present a passbook for each transaction. This passbook could be presented in person or mailed in with a deposit or withdrawal. Since consumer savings accounts provide a substantial portion of the bank's funds for other services, the bank has embarked on a program of creating new and innovative features for existing accounts and adding new types of accounts, all designed to lure more of the customers' funds for a longer period.

Some of these new savings products are: money market accounts, NOW accounts (negotiable order of withdrawal), and certificates of deposit with lower than normal deposit requirements.

Because of the attractiveness and popularity of individual retirement accounts (IRAs), the bank has developed a number of products which can serve as investment instruments for these long-term deposit accounts.

Consumer Loans

Much of the bank's traditional revenue comes from its consumer loan and credit operations, which include the traditional mortgage, auto, and home improvement loans. In recent years, increased competition and the development of new products have put increased pressure on these revenue mechanisms. The current mortgage environment has changed from fixed-rate fixed-term mortgages to current flexible-rate flexible-term instruments. Additionally, the new home equity loans and equity access accounts (which in effect represent second mortgages) greatly increase the risk in this area. The increased use of credit cards and consumer credit lines has also made it imperative that the bank have means for assessing its exposure in these areas.

The fluctuating and currently declining interest rate environment also puts pressure on the bank to ensure that its income sources keep pace with its expenses (primarily interest owed). Changing tax laws, which reflect tighter controls on interest expenses, also affect bank income and expense figures and mandate that the bank track its commitments in this area very closely.

Customer Service

In keeping with its new emphasis on improved customer service and its new customer orientation, the bank would like to produce consolidated customer statements. Currently a customer receives a statement for each account type. Aside from being confusing and annoying to the customer, it is costly for the bank. Each statement must detail all activity against that account during the preceding statement period. The separate statements, however, do not give customers a complete picture of their banking relationships. A consolidated statement would provide a summary balance for each account relationship, including credit card and loan accounts, and would allow customers to see their entire financial position at a glance. In addition the statement would continue to provide the detailed account transaction activity as before.

VERYLARGE FULL SERVICE BROKERAGE COMPANY

Background

The Verylarge Full Service Brokerage Company was established over 100 years ago as a partnership. Its primary orientation and the bulk of its operations, then as now, is in the area of retail brokerage. Retail brokerage deals primarily with individual investors and small businesses. The primary activities in this area are the execution of buy and sell (trades) orders for securities and other financial instruments on behalf of its customers and the management of the investment portfolios of these customers. To this end it established a network of storefront branch offices, each staffed with a sales force and a back office processing area.

In addition to the execution of securities trades, the sales force is empowered to sell to its clients any number of financial products, which include shares in mutual funds, specialized investment partnerships, commodities (mainly farm products and precious metals), retirement and annuity products, insurance products, investment newsletters, etc.

The firm, through its sales force, acts as an intermediary rather than as a principal in these securities transactions for the customer. That is, it is not an active party to the trade or sale, but rather arranges the buy or sell and takes a commission based upon a formula which takes into account the price of the securities or other products, and the number of units traded.

These trades may be effected in one of the national, regional, or international markets (such as the New York or American Stock Exchanges, the Boston, Pacific, or Montreal Stock Exchanges, or the London or Tokyo Stock Exchanges).

For nonsecurities transactions, those where the firm is selling its own product (its own mutual fund, newsletters, or insurance policies written against its own insurance subsidiary), or where the firm is buying for or selling from its own securities inventory, the firm is also an active party to the trade.

The Sales Force

Company "sales" are handled by a combination of commissioned salespeople and salaried customer service representatives. All securities salespeople must be licensed by the Securities and Exchange Commission and in some cases by other regulatory or governmental units. In order to receive this license they must take a rigorous course in securities operations, rules, laws, and regulations. Upon completion of this course they are licensed and are then "registered" to execute trades on behalf of their customers and provide investment advice.

Each salesperson is responsible for signing up and servicing his or her own customers. Before customers can trade, however, they must establish an account with the firm. Because of regulatory requirements and in order to facilitate accounting, each type of trading activity is usually handled through a different type of account. Salespeople are compensated based upon the number of accounts they open and control and upon the level of activity in each account. Although most salespeople are, technically speaking, employees of the firm, many consider themselves to be, and thus operate as, independent agents who "work at" rather than "for" the firms they represent.

The customer service representatives may or may not be licensed to trade, and in many cases may be salespeople "in-training." Each customer service representative is normally assigned to one or more salespeople and is responsible for ensuring that the paperwork for account maintenance is properly handled and that trade-related forms are properly routed. Customer service representatives also handle routine customer inquiries. They may not give securities or other trading advice (even though they may be licensed to do so) and may not execute any actual trades. In contrast to the salespeople, customer service representatives are full-time, salaried employees of the firm. Many sales representatives also receive bonuses based upon how well "their" salespeople or "their" office does.

Types of Accounts

Each type of customer account is tailored to a particular type of trading or investment activity. There are two basic types of accounts: cash and margin.

The most usual account is the cash account, which permits the customer to trade on a cash basis, as opposed to credit. That is, all buy transactions must be settled in full, in cash.

Margin accounts by contrast are accounts where the firm allows the customer to put up only a part of the value of the securities purchased, in cash, with the company providing the remainder of the value in the form of a loan to the customer. Thus the customer could buy $10,000 dollars worth of securities for as little as $5000 or $6000 in actual cash, with the company financing the remainder of the cost. The firm charges the customer interest on the money loaned. For margin accounts, the firm must, by law, maintain a fixed ratio between the actual market value of the equity and the customer's actual investment. If that ratio falls below the regulatory amount, the firm must obtain further funds from the customer within a fixed number of business days.

The firm will normally open up a separate account for each type of security, or investment in which the customer can trade: equity (stock), debt (bonds), commodities, currency, money market funds, and both equity and debt mutual funds. In addition the customer may also have separate accounts for any special products which the firm may offer, such as investment partnerships, tax shelters, retirement funds, annuity funds, etc. A customer may also open and control separate accounts of the same type in the name of family members, for special business purposes, etc.

Customer Accounting

The firm treats customer accounts in much the same manner as commercial and savings bank accounts are treated. That is, they are summary accounts, and all transaction details are recorded on a separate basis, with final totals being kept in the account itself. The firm will send customers an "advice" which notifies them of the completion of each trade, each interest or dividend payment, or other transaction; on a monthly basis the firm sends customers a statement which details all activity, starting and ending balances, and their position in all securities or instruments which have been traded in the name of the account since the last statement.

Because of the high volatility of trades in many accounts, for security reasons, and to reduce the amount of paperwork, most firms offer to hold customer securities on their own premises or in independent "depositories." In these cases, although customers retain full ownership rights to the securities, they do not normally take physical possession. The brokerage firm in these cases will receive and distribute to the clients all dividend or interest payments earned by the securities, and these payments will be reflected on the monthly statements.

Because they physically hold the documents for the customer, the firm must have systems which allow it to track these securities inventories. Normally the firm must track inventory positions in two ways: (a) they must know at all times which customers hold what securities (for portfolio valuation purposes) and (b) which securities are held by each customer (for margin calculation, and interest and dividend payment purposes).

Trade Processing

Each trade is in effect an order from a customer to buy or sell some security or other instrument. The order may be "at market" which means at whatever the price is at that time, or it may be an order with limits or restrictions. These limits or restrictions may be with respect to either price, time, or a combination of both. That trade is forwarded to a trader specialist who contacts an opposite number and "executes" the trade.

The customer then normally has five business days to "settle" the trade. Settlement consists of either delivering payment for that which was bought or delivering the document which represents that which was sold. The actual ownership of the security must be changed, and the company whose shares it represents must be notified of that change. The firm must then execute the physical (or book record) transfer of ownership and possession. Because of the monetary value of these securities, there are strict accounting and regulatory controls placed upon the movement of these documents.

At each stage in the process numerous reports and records are maintained to ensure that all regulations and procedures are complied with and that the financial and inventory accounts reflect the current status of the affected accounts.

Glossary of Terms and Concepts

Activity That set of tasks which are organized and broken down into a set of procedures to accomplish a specific goal. The distinction between a subfunction and an activity is as much a matter of interpretation as it is a matter of scope.

Analysis The separation of an intellectual or substantial whole into its constituent parts for individual study. The stated findings of such a separation or determination.

Application The specific set of activities under analysis. An application may consist of one or more activities within a functional area, or it may include all activities within a functional area. In some cases the application may cross functional areas. In some firms an application is synonymous with a system.

Attribute An aspect, quality, or characteristic of either an entity or a relationship which describes it. An attribute may be a physical characteristic, such as size, weight, or color, or a locational attribute, such as place of residence or place of birth. It may be a quality such as level of a particular skill, educational degree achieved, or the dollar value of the items represented by the order.

Data analysis That process by which the data requirements of a functional area are identified, element by element. Each data element is defined from a business sense, its ownership is identified, and users and sources of that data are identified. These data elements are grouped into records, and a data structure is created which indicates the data dependencies.

Data dictionary An automated tool for collecting and organizing the detailed information about system components. Data dictionaries maintain facilities to document data elements, records, programs, systems, files, users, and other system components. A dictionary will also have facilities to cross-reference all system components to each other.

Data element The lowest unit of meaningful information in an automated file or on a document. A data element may consist of numbers, letters, or a combination of both.

Entity Any real person, place, or thing, or logical person, place, or thing which can be definitively described, and which is of immediate and/or ongoing interest to the firm as a whole or to some aspect of the firm. An entity may also be an idea, concept, or convenience.

Entity set All known or suspected variants of the singular entities which make up the global set. In the entity-relationship model, the entity set is treated as if it were synonymous with the individual entities which comprise it. That is, the set is treated as if each of its component entities is defined and behaves in a similar manner.

File A group of records, in automated or document form, which relate to the same subject and which are used and manipulated in the same manner.

Function A series of related activities, involving one or more entities, performed for the direct or indirect purpose of fulfilling one or more missions or objectives of the firm, generating revenue for the firm, servicing the customers of the firm, producing the products and services of the firm, or managing, administering, monitoring, recording, or reporting on the activities, states, or conditions of the entities of the firm.

Interview A formal face-to-face meeting, especially, one arranged for the assessment of the qualifications of an applicant, as for employment or admission. . . . A conversation, as one conducted by a reporter, in which facts, or statements are elicited from another.

Method A means or manner of procedure, a regular and systematic way of accomplishing something. An orderly and systematic arrangement. Procedures according to a detailed, logically ordered plan.

Methodology The system of principles, practices, and procedures applied to a specific branch of knowledge.

Model A representation, either graphic, narrative, or a combination of both, of a physical or conceptual environment. A model must identify the major components of the environment, describe those components in terms of their major attributes, and depict the relationships between the components and the conditions under which the components exist and interact with each other.

Plan That sequence of activities which are to be followed. A plan states each task, the estimated time to complete it, the persons assigned to perform it, and any task-to-task dependencies. Plans are updated on a periodic basis with actual results, and new estimates are determined. At any point, the plan should reflect actual progress and remaining work.

Procedure The specific steps which must be followed in order to accomplish a specific task or activity.

Process A sequence of related activities, or it may be a sequence of related tasks which make up an activity. These activities or tasks are usually interdependent, and there is a well-defined flow from one activity to another or from one task to another.

Program A sequence of instructions which may be followed by a computer to perform a specific task or tasks.

Record A group of one or more data elements which are stored together and which represent information which relates to a common topic. A record may be automated, or it may be a business document.

Relationship An association, linkage, or connection, either real or suspected, between entities of the same or different set which describes their interaction, the dependence of one upon the other, or their mutual interdependence.

Security The protection of the firm's records and resources from unauthorized access, modification, or other interference includes an analysis of ownership, access, modification, use, and a determination of what protective or restrictive measures must be taken to ensure adequate protection of the firm's files.

Standards The rules which must be followed in order to accomplish a specific activity or task. Standards are established to ensure that all work is performed in a uniform manner.

System A group of interacting, interrelated, or interdependent (business functions, processes, activities or) elements forming a complex whole . . . . a functionally related group of (business functions, processes, activities or) elements, for instance, a network of structures and channels, as for communications, travel, or distribution.

Systems analyst One who engages in the study of, and separation of, a group of interacting, interrelated, or interdependent (business functions, processes, activities or) elements forming a complex whole into its constituent parts for individual study.

Task The lowest unit of discrete work which can be identified. An activity may be composed of many tasks. Tasks are highly repetitive, highly formalized, and rigidly defined.

Users Business personnel in other areas of the firm who manage, supervise, or perform the direct and indirect operational, managerial, and administrative tasks of the firm. Users provide the impetus for the development of these systems, in many cases they fund the development and implementation process and provide for their ongoing operation, and in all cases they supply the policies, guidelines, business requirements, specifications, and background information about the particular area to be systematized and automated.

INDEX

ABOUT THE AUTHOR

Martin Modell holds an MBA in General Management from the New York Institute of Technology. With over 25 years of general data-processing experience, he has been directly involved in all phases of database technology and management since 1970. A contributing reviewer on database topics for the ACM Computing Reviews, he has also been a participant and speaker at many national and regional database technical and design conferences. Currently a senior information systems architect and data analysis and data modeling consultant for Unisys Corporation, he is vice chairman of the steering committee for the International Conferences on Entity-Relationship Approach and international coordinator for the Entity-Relationship User's Groups. Mr. Modell is a member of the ER Institute and a founding and guiding member of the International committee formed to develop Entity Relationship Approach and Data Modeling standards.